Introduction to stellar astrophysics

Volume 1
Basic stellar observations and data

Introduction to stellar astrophysics

Volume 1
Basic stellar observations and data

Erika Böhm-Vitense
University of Washington

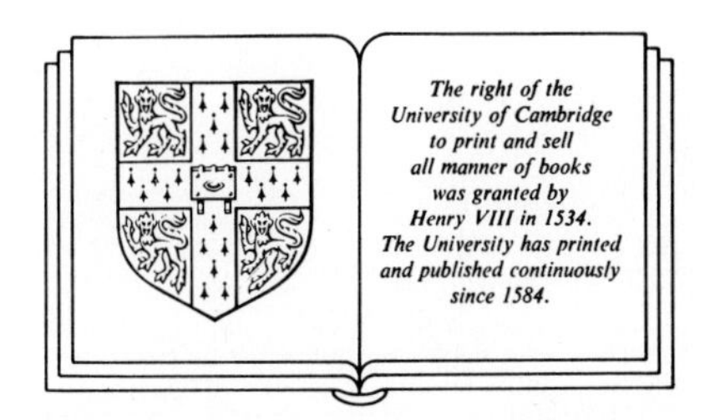

CAMBRIDGE UNIVERSITY PRESS

Cambridge

New York   Port Chester   Melbourne   Sydney

Published by the Press Syndicate of the University of Cambridge
The Pitt Building, Trumpington Street, Cambridge CB2 1RP
40 West 20th Street, New York, NY 10011, USA
10 Stamford Road, Oakleigh, Melbourne 3166, Australia

First published 1989

Printed in Great Britain at the University Press, Cambridge

British Library cataloguing in publication data

Böhm-Vitense, Erika, 1923-
Introduction to stellar astrophysics.
Vol. 1: Basic stellar observations and data
1. Stars
I. Title
523.8

Library of Congress cataloguing in publication data

Böhm-Vitense, E.
Introduction to stellar astrophysics/Erika Böhm-Vitense.
p. cm.
Includes index.
Contents: v. 1. Basic stellar observations and data.
ISBN 0-521-34402-6 (v. 1). ISBN 0-521-34869-2 (pbk. : v. 1)
1. Stars. 2. Astrophysics. I. Title.
QB801.B64 1989
523.8—dc19 88-20310 CIP

ISBN 0 521 34402 6 hardcovers
ISBN 0 521 34869 2 paperback

TM

Contents

Preface

The topic of this volume is stellar astronomy or more accurately stellar astrophysics. We call it astrophysics because all our knowledge about stars is based on the application of the laws of physics to the stars. We want to find out how big the stars are, how much mass they have, what material they are made of, how hot they are, how they evolve in time, and how they are distributed in space. The last question does not strictly belong to the field of stellar astrophysics but knowledge of stellar structure and evolution will provide a means by which to determine their distances. There are also important correlations, for instance, between the location and motion of the stars in our Galaxy and their physical properties.

In Volume 1, we shall be concerned mainly with finding out about the global properties of stars, such as brightnesses, colors, masses and radii. Brightnesses and colors can be measured directly for all stars, for masses and radii we have to study binaries. Parallax measurements can give us distances to nearby stars. We shall first discuss the majority of stars which we call normal stars. In the latter parts of this volume we shall also look at stars which seem to be different, the so-called 'peculiar' stars.

How can we get information, for instance, about the physical properties of the stars such as their temperatures, pressures, and chemical compositions? For most of the stars we have to get this information from that tiny little bit of light which we can receive. Only the sun gives us a lot of light to study, but even for this nearby star we get little more than just the light. For all the other stars it is only this tiny little dot of light that has to tell us all about the properties of the star from which it comes. It will be the topic of Volume 2 to study how we can do this.

In Volume 3 we will use theoretical considerations and the known laws of physics to derive the properties of stellar interiors, and discuss stellar evolution, as well as the origin of the chemical elements. We shall see which observations can tell us something about the structure of the insides of the stars.

The critical reader might wonder whether we are justified in applying the laws of physics, as we know them from our experiments on Earth, to the stars. How do we know whether the same laws apply to the stars? We really do not know. We can, however, try to understand the stellar observations, assuming that the same laws of physics hold. As long as we can do that and get sensible results, and if we can predict successfully the results of further observations we can feel that we are on the right track. Our space travel within the solar system so far has obeyed the laws of physics as we know them. Once we come to a point where we can see that the observations clearly contradict our earthly laws of physics we will have to make corrections. So far this has not been necessary, except that some refinements have been made. These refinements are, however, believed or proven to hold also in experiments on Earth, except that some effects are too small to be measurable in laboratory experiments. Our laboratories are too small to measure some effects such as the bending of a light beam in a gravitational field as predicted by the theory of general relativity. To measure these effects we need the largest laboratory which we can get, the universe.

The three volumes were written for students in their junior and senior years. They should be understandable by the educated layman with some basic knowledge of physics and mathematics.

I apologize for not giving all the references to the authors who have contributed to our present day knowledge of stars as described here. They would fill a large volume by themselves. I quote at the end only a few textbooks which also describe some of the observational results discussed here, and I list a number of reference books and tables which also give basic data about stars. I do list all the references from which figures and tables were used. Frequently these references are also textbooks or review articles which will lead the reader to a more detailed discussion than is possible in this short volume.

I am very much indebted to Dr G. Wallerstein for detecting several errors and weaknesses in the manuscript, and to Dr R. Schommer for a critical reading of the whole manuscript and for many helpful comments.

I am also very grateful for a JILA fellowship which permitted me to write most of this book. I acknowledge especially the help of the JILA secretaries with the typing of the manuscript.

1

Positions of stars

1.1 The coordinate system

If we want to study stars, the first thing to look at might be their positions in the sky. This in itself does not tell us much about the nature of the stars, but it is very helpful when we want to find a particular star or a group of stars in the sky. We have to have a reference point with respect to which we can describe the position of the star in which we are interested. We all know or have at least heard about the constellations of stars which in earlier times were extremely helpful in describing the positions of stars with respect to a given star in a particular constellation. We still name the brightest stars according to the constellations in which they are found, but we like to have a more general description of the positions. When looking at the sky we can measure the positions only as projected against the sphere of the sky, i.e., against a two-dimensional surface. We can therefore describe the positions of the stars by two quantities. Since the surface against which we measure the positions is a sphere, we use spherical polar coordinates. Since our telescopes are fixed on the Earth, we use a coordinate system which is fixed with respect to the Earth. The Earth is rotating, but we do not want to have a rotating coordinate system, which would cause many problems. We keep the coordinate system fixed in space. The equatorial plane of our spherical polar coordinate system is identical with the equatorial plane of the Earth, which means that the equatorial plane is perpendicular to the rotation axis of the Earth. Unfortunately, the direction of the rotation axis of the Earth is not fixed in space, but due to the gravitational forces of the sun and the moon on the Earth, the Earth's axis of rotation is precessing, i.e., describes approximately a cone around an axis fixed in the Earth. This causes our reference plane also to precess, which means that the coordinates of the stars are changing in time because the axes of the coordinate system are changing in time, not because the stars are changing their positions. Of course, the stars are also moving in space but that leads to much smaller changes in the coordinates than does the precession of the Earth's rotation axis.

The coordinates which the astronomers use are the right ascension α and the declination δ. The right ascension α corresponds to the longitude, which we use on the Earth's surface to describe the position of a particular place, and the declination δ corresponds to the latitude which we use on the surface of the Earth, see Fig. 1.1. As we know from the Earth, we still have to define the meridian which we call longitude zero. On Earth this is defined as the meridian which goes through Greenwich. On the celestial sphere we also have to define a meridian through a given point as being longitude or right ascension zero. We could define the position of a given star as right ascension zero, but then that star might turn out to move in space and then that coordinate system would move with this arbitrarily chosen star. We could choose the position of a very distant object, for instance, the position of a quasar. Even a large space motion of such a distant object would not change its position measurably. At the time when the coordinate system was defined the quasars were not known and the distances of other astronomical objects were not known either. The zero point for the right ascension was therefore defined by the direction of a line, namely the line given by the intersection of two planes, the equatorial plane of the Earth and the orbital plane of the Earth around the sun, the ecliptic, see Fig. 1.2. As the orientation of the equatorial plane changes with time because of the precession of the Earth's rotation axis, the direction of the line of intersection of the ecliptic and the equatorial plane also changes with time which means the zero point for the right ascension also changes with time. So the coordinates for all the stars change with time in a way which can be computed from the known motion of the Earth's rotation axis. The right ascension is measured in hours,

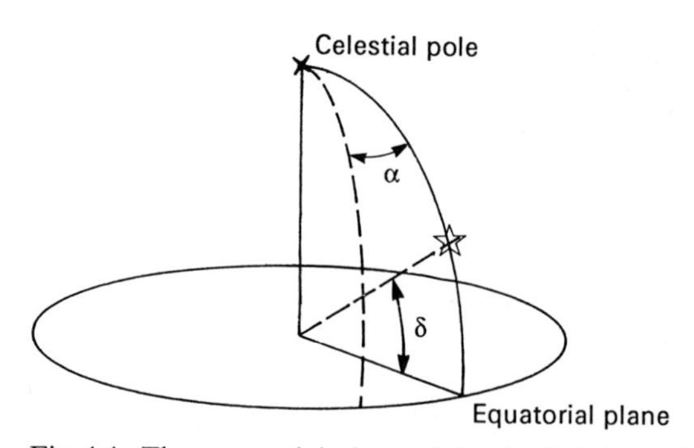

Fig. 1.1. The equatorial plane of the Earth defines the plane for the celestial polar coordinate system, which describes the positions of the stars by right ascension α and the declination δ.

minutes, and seconds. 24 hours correspond to 360 degrees. The right ascension gives the siderial time when the star has its greatest altitude above the horizon. The declination δ is measured in degrees, $-90° < \delta < +90°$.

There are catalogues which give the coordinates of the stars for a given year, we then have to calculate the corrections to the coordinates for the time, when we want to observe the object. The equations for computing these corrections can be found in Smart's textbook on spherical astronomy (1977). Tables for the corrections are given by Allen (1982).

Catalogues with positions of stars for the year 1855 are, for example, the 'Bonner Durchmusterung' (BD), and for 1900 the 'Henry Draper' Catalogue (HD). Stellar positions for the year 1950 are given in the catalogue of the Smithsonian Astrophysical Observatory (S.A.O.).

1.2 Direction of the Earth's rotation axis

From the previous discussion it is apparent that we have to know how the position, or better, the direction of the Earth's rotation axis changes in time. How can we determine this direction? The best way is to take a long exposure photograph of the sky with a fixed orientation of the telescope, preferably close to the direction of the north polar star. Because of the Earth's rotation, which will cause the telescope to change its orientation in space, the stellar positions will apparently move in circles around the direction of the rotation axis during the course of a day, see Fig. 1.3.

Repeated observations of this kind permit a determination of the changing direction of the Earth's rotation axis.

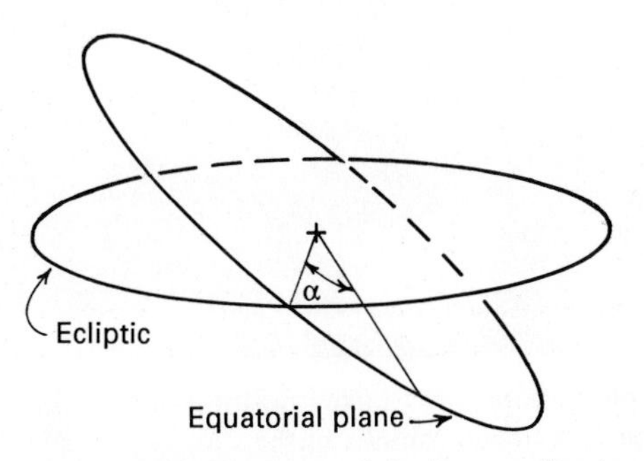

Fig. 1.2. The direction of the intersection between the equatorial plane and the plane of the ecliptic defines the meridian for the zero point of the right ascension. When the orientation of the equatorial plane changes, the position of the zero point meridian changes.

1.3 Visibility of the sky

From every point on the surface of the Earth we can see only a fraction of the sky which is determined by our horizon and by the rotation of the Earth, as illustrated by Fig. 1.4. Suppose the observer is at point P on the Earth early in the morning. The plane of his horizon is indicated by the solid line. He can observe everything which is above his horizon. Of course, he will only see the stars if the sun is not shining on his side of the Earth. Twelve hours later the observer will be at point P' because of the Earth's rotation

Fig. 1.3. A long-exposure photograph with a fixed position of the telescope pointing towards the North Pole. The positions of the stars describe circles in the sky with the centers of the circles showing the direction of the rotation axis of the Earth. The lengths of the circle segments seen in the photograph are determined by the duration of the exposure. A 12-hour exposure would give a half-circle. (From Abell 1982.)

around the axis ω. The plane of his horizon is now indicated by the dashed line. He can see only what is above this plane. The whole cone, which is cut out by the rotating plane of the horizon is excluded from his view. Only observers at the equator have a chance of seeing the whole celestial sphere during one day, however, they will only be able to see all of the stars during the course of one year, because the sun always illuminates about half of the sky.

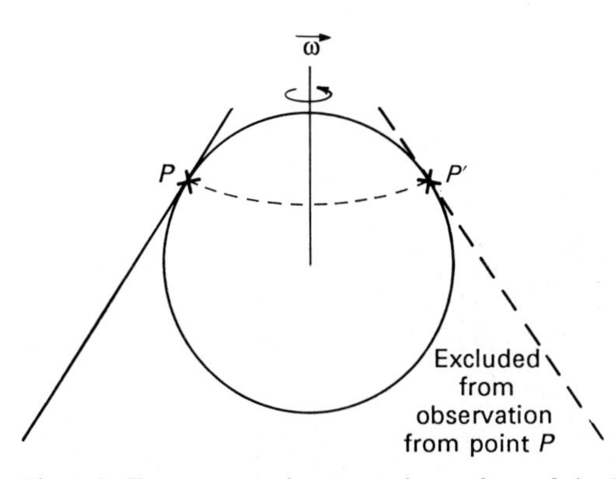

Fig. 1.4. From any point P on the surface of the Earth, a cone is excluded from observation, except for points on the equator.

2

Proper motions of stars

In the previous chapter we have seen that the coordinates of the stars change with time because the coordinate system, which is defined by the rotation axis of the Earth, changes with time. The coordinates of the stars may also change because the stars themselves move in space. Only the motions perpendicular to the direction of the line of sight will actually give a change in the coordinates, see Fig. 2.1. Motions along the line of sight will change the distance but not the coordinates. The motions perpendicular to the line of sight are called proper motions, because they give coordinate changes which are due to the star's proper and not to the Earth's rotation. Velocities along the line of sight are called radial velocities because they go in the direction of the radius of a sphere around the observer. Proper motions are measured by the changes in their right ascension and declination, which are angles. The proper motions are therefore measured in sec or arcsec per year, while the radial velocities are measured by means of the Doppler shift, see Section 9.2, which gives the velocities in km s^{-1}. It would be difficult to give proper motions in km/s because the relation between proper motions in arcsec and in km s^{-1} depends on the distance of the star: A given velocity

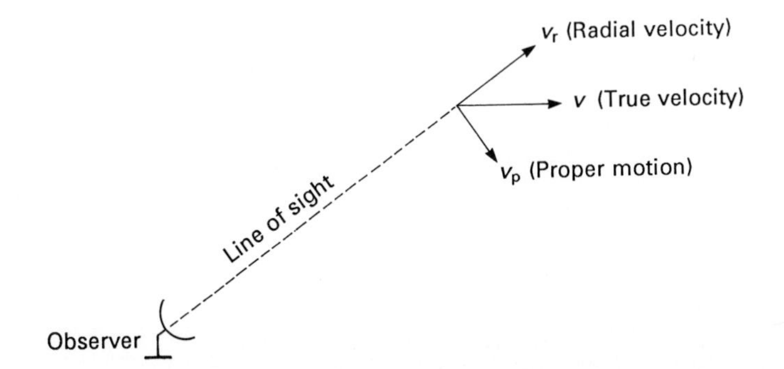

Fig. 2.1. Only motions in the direction perpendicular to the line of sight change the position of the star in the sky, i.e., the point of the projection of the star against the background sphere. Radial velocities do not change the coordinates of the star, only its distance.

of a star in a direction perpendicular to the line of sight will lead to a relatively larger change in position, i.e., to a relatively large proper motion, if the star is nearby and only to a very small change in position if the star is far away, see Fig. 2.2. In fact, proper motion studies are often used to find nearby stars.

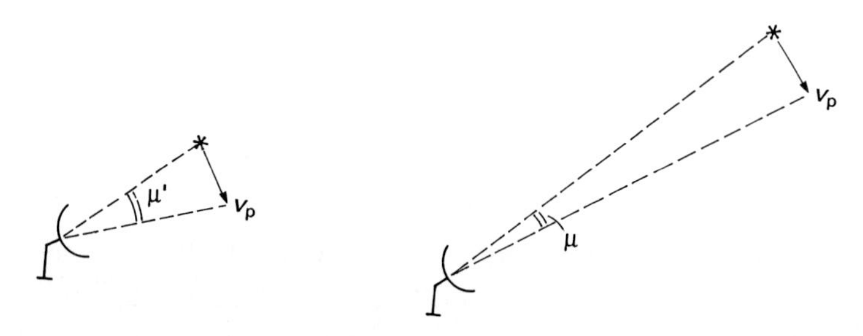

Fig. 2.2. For a given velocity v_p perpendicular to the line of sight the proper motion in arcsec (μ') is larger for a nearby star than for a more distant star (μ).

3

Distances of nearby stars

3.1 The distance of the sun

The distances to nearby objects on Earth are determined by measuring how often a stick of standard length, for instance, a meter stick, fits in between the two objects whose distance we want to measure. For larger distances, this very often does not work. For instance, in a mountain area it would be impossible to measure the distances of two mountain tops in this way.

Our eyes make rough distance determinations without using a meter stick. Our eyes actually use the so-called method of triangulation. For triangulation we observe a given object from two different points whose distance we know, for instance, by measuring with a meter stick.

From two observing points A and B the observed object C will appear projected against the background at different positions E and D, see Figure 3.1. For a nearby object there will be a large angle γ between the projection points, for an object further away the angle will be smaller, see Fig. 3.2. The relation between the angle γ measured from the two points and the distance to the object is given by

$$\sin\left(\frac{\gamma}{2}\right) = a/(2d), \tag{3.1}$$

where a is the distance between the two observing points, A and B, and d is the distance to the object, C or C', from the center of the two observing points,

For large distances we can set $\sin\gamma = \gamma$ if γ is measured in radians. (3.1) can then be replaced by

$$\gamma = a/d \tag{3.2}$$

from which d can be determined if a and γ have been measured. For us our two eyes serve as the two observing points, a is then the distance between the eyes.

From (3.2) it is obvious that we can measure larger distances if the baseline a is large, because there is a limit to the size of the angle γ which we can still measure. On Earth there is a limit to the length of the baseline a, this limit is determined by the diameter of the Earth. It turns out that this baseline is not large enough to measure even the distance to the sun accurately. We can, however, measure distances to nearby asteroids this way and then use Kepler's third law to determine the distance to the sun. We can now also determine the distance to Venus by radar measurements and then again use Kepler's third law to derive the distance to the sun. The method works as fellows: Kepler's third law states that the squares of the orbital periods of the planets are proportional to the third powers of the semi-major axis b of their orbits around the sun, or

$$P^2/b^3 = \text{const.} = A. \tag{3.3}$$

For two planets, for instance, the Earth and Venus (or an asteroid) this tells us that

$$P(\text{Venus})^2/P(\text{Earth})^2 = b(\text{Venus})^3/b(\text{Earth})^3, \tag{3.3a}$$

where b(Venus) and b(Earth) are the semi-major orbital axis of Venus and Earth. P(Earth) is one year and P(Venus) is the orbital period of Venus, which is 224.7 days. Equation (3.3a) is one equation for the two semi-major axes of Venus and Earth. If we had one more equation we could determine

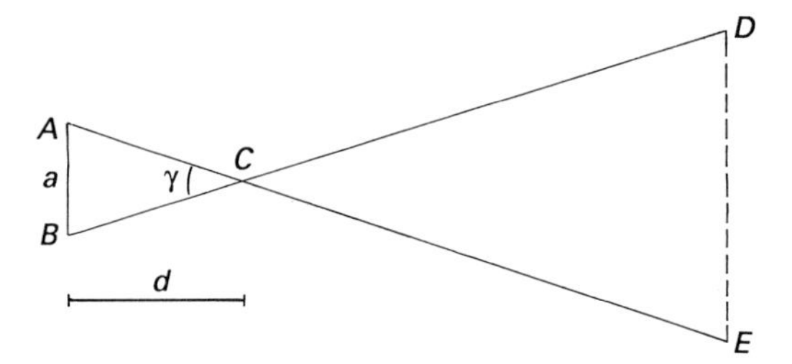

Fig. 3.1. From two observing points A and B, distance a apart, the object C is projected against the background at different points D and E. The angle γ at which the object appears from the observing points A and B is given by (3.1); see text.

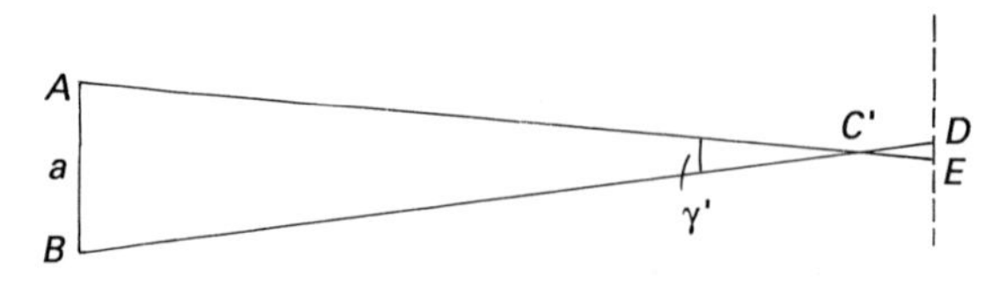

Fig. 3.2. For a distant object C' the angle γ' at which the object appears from the observing points A and B is smaller than for a nearby object C.

both semi-major axes. The second equation is provided by a measurement of the distance between Venus and Earth at their nearest approach when it can best be measured, see Fig. 3.3. In order to demonstrate the principle, we approximate both orbits by circles (the ellipticities are actually quite small). From Fig. 3.3 we see that then the distance Venus–Earth d is

$$d = b(\text{Venus}) - b(\text{Earth}). \tag{3.4}$$

If we measure d, then (3.4) provides the second equation needed to determine both b(Venus) and b(Earth). Of course, b(Earth) is the distance Earth–sun, which we want to determine, and which is generally called one astronomical unit or abbreviated 1 au. The distance Venus–sun is only 0.72 au. At closest approach the distance Venus–Earth is therefore only 0.28 au, see Fig. 3.3. This distance is small enough to be measured by radar. Since the radar signal travels with the speed of light c, it takes the radar signal a time t to travel to Venus and back which is given by

$$t = 2\,d/c. \tag{3.5}$$

The time t can be measured and d can be determined from (3.5). Using (3.3a) and (3.4) we then derive the distance Earth–sun to be

$$1\ \text{au} = 1.49 \times 10^{13}\ \text{cm}.$$

Taking into account the ellipticities of the orbits will complicate the mathematics but does not change the principle.

3.2 Trigonometric parallaxes of stars

Once we know the orbital diameter of the Earth we can use this length as the baseline, a, for further triangulation. If we make one observation on day 1 and the second observation half a year later, we have changed our position in space by the orbital diameter of the Earth. It is not necessary to make the observations from the two observing points at the same time.

From Fig. 3.4 we infer that for a star at the ecliptic pole we will see a change

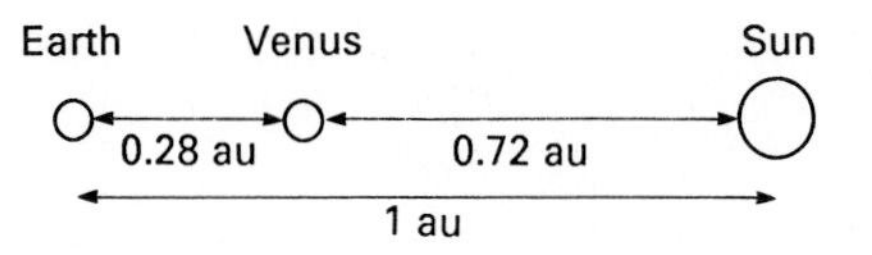

Fig. 3.3. The smallest distance Earth–Venus is only 0.28 times the distance Earth–sun.

of position during half a year which is given by the parallax angle $2\pi = 2\,\mathrm{au}/d$ or

$$\pi = 1\,\mathrm{au}/d,$$

where d is the distance sun–star. Knowing the size of one au and measuring π will give us the distance to the star, provided the angle π is large enough to be measured. Since the distances of the stars are so large it is convenient to introduce a new unit of length to measure stellar distances. This unit is called the parsec (pc) and is the distance at which a star would be if it had a parallax angle π equal to one arcsec. From the geometry in Fig. 3.3 we derive that such a star would have to be at a distance d equal to 3.08×10^{18} cm. The light would need about three years to travel this distance. Actually, there are no stars which have a parallax angle as large as 1 arcsec. The closest star, Proxima Centauri, has a parallax angle of 0.76 arcsec. We can only measure parallax angles larger than about 0.05 arcsec with reasonable accuracy, which means we can determine distances out to about 20 pc with fairly good accuracy; beyond that trigonometric parallaxes cannot be of much help. As our measuring instruments improve we may be able to measure parallaxes down to 0.02 arcsec. We are lucky that we find about 200 stars within 20 pc distance, bright enough to be the basis for the distance determinations. There are several thousand very faint stars within 20 pc distance.

In Fig. 3.4 we have demonstrated only the case for a star at the ecliptic pole which during the course of one year describes a circle projected against the background sky due to the orbit of the Earth around the sun. If the stars are not near the ecliptic pole they will, of course, also show parallax motions against the background sky but their 'parallax orbits' will not be circles but ellipses. For the major axes of these ellipses, the same equation holds as for the 'orbital' radii of the stars at the ecliptic poles.

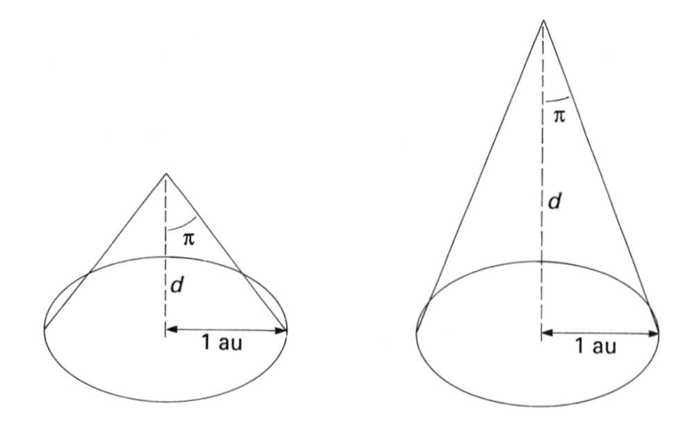

Fig. 3.4. The parallax angle π at which a star is seen when measured at six-month intervals is a measure of its distance. The further away the star, the smaller is π.

4

The brightnesses of the stars

4.1 The apparent magnitudes

The brightness of a star is a quantity which can be measured easily, at least in a qualitative way, by comparing the brightnesses of different stars. Even with the naked eye we can see whether star A is brighter than star B or vice versa. Ancient astronomers called the brightest stars first magnitude stars, the next fainter ones were second magnitude stars and so on. These magnitudes were determined by the sensitivity of the eyes because all observations were made by the naked eye. The sensitivity of the eye is logarithmic to enable us to see a large range of light intensity. The magnitude scale is therefore a logarithmic one. A given difference in brightness corresponds to a given factor in the amount of energy received.

When astronomers learned to make quantitative intensity measurements, they did not want to part with the well-established magnitude scale. They therefore put the magnitude scale on a quantitative basis. Since it is a logarithmic scale, as we saw, they found the old scale could best be represented by defining

$$\Delta m_v = -2.5 \cdot \Delta \log I_v, \tag{4.1}$$

where Δm_v is the difference is magnitude as seen by the eye, i.e., by visual observation, and I_v is the intensity received with an instrument which has a color sensitivity similar to our eye. For two stars A and B, we then derive

$$m_v(A) - m_v(B) = -2.5[\log I_v(A) - \log I_v(B)]. \tag{4.2a}$$

The minus sign on the right-hand side takes care of the fact that the magnitudes become larger when the stars become fainter. Remember the stars of first magnitude are brighter than those of second magnitude!

If we actually want to determine the magnitude of star B, we have to know the magnitude of star A:

$$m_v(B) = m_v(A) + 2.5[\log I_v(A) - \log I_v(B)]. \tag{4.2b}$$

The magnitude scale was originally defined by the north polar sequence of stars. It is now defined by a number of stars measured accurately by Johnson and Morgan in 1953. For all practical purposes, we can say that the magnitude scale is defined by assigning the magnitude 0 to the star α Lyrae, also called Vega. So if star A is Vega, then $m_v(A) = 0$ and

$$m_v(B) = 2.5[\log I_v(\text{Vega}) - \log I_v(B)]. \tag{4.2c}$$

(Actually $m_v(\text{Vega}) = 0.02 \pm 0.01$, but we do not here worry about this small difference.) In practice, we compare the brightness of all stars with that of Vega. If a star is fainter than Vega, then $m_v > 0$, if a star is brighter than Vega, then it has $m_v < 0$. There are some stars brighter than Vega, for instance, Sirius, these stars then have *negative* magnitudes. Sirius has $m_v = -1.6$.

We have always indicated these magnitudes with a *lower case m* because they refer to the brightnesses as we see them. They are called *apparent brightnesses* or *apparent magnitudes*. These apparent brightnesses have to be distinguished from the intrinsic brightnesses, called *absolute brightnesses* or *absolute magnitudes* which are designated by a *capital M* (see Section 4.6).

4.2 The colors of the stars

We have emphasized that the apparent brightnesses or magnitudes discussed above refer to visual observations. They compare the brightnesses as seen with the eye. They are called the visual magnitudes.

Quantitative measurements are now made either with photographic plates or more frequently with photoelectric instruments. These measuring devices have a different sensitivity from our eyes. Originally photographic plates were mainly sensitive to blue light. With photographic plates we compared the brightness of the stars in the blue wavelength band. A star, which in the visual has the same brightness as Vega and therefore has $m_v = 0$, may actually be brighter than Vega in the blue, if it is a blue star, so its blue magnitude will be $m_B < 0$. A given star will generally have different magnitudes for different wavelength bands. If it is more blue than Vega, its blue magnitude will be smaller than its visual magnitude, a star that has relatively more energy in the red as compared to Vega will have a larger blue magnitude than visual magnitude. The difference in magnitude of a given star for different wavelength bands tells us something about the color of the star. For $m_B > m_v$ the star is *fainter* in the blue, i.e., the star looks more red than Vega. *For Vega all apparent magnitudes are zero* by definition, see (4.2c) which holds actually for all wavelength bands. (This does *not* mean that Vega has the same brightness at all wavelength bands.)

The most widely used system of apparent magnitudes is the so-called UBV system which measures apparent magnitudes in the ultraviolet, U, referring to ~3600 Å, in the blue, B, referring to ~4300 Å, and in the visual, V, referring to ~5500 Å. The sensitivity functions of the measuring instruments are shown in Fig. 4.1. There are many different magnitude systems in use, almost as many as there are astronomers who measure magnitudes. They all serve their special purposes. The UBV system has the largest number of measurements. Each band uses a large fraction of the spectrum (1000 Å) and can therefore be used to measure magnitudes of rather faint stars. In the following discussion we will therefore use only this system.

The difference in the ultraviolet magnitudes and blue magnitudes for a given star is abbreviated by

$$\mathrm{U}-\mathrm{B}=m_{\mathrm{u}}-m_{\mathrm{B}}, \tag{4.3}$$

and

$$\mathrm{B}-\mathrm{V}=m_{\mathrm{B}}-m_{\mathrm{v}}. \tag{4.4}$$

According to our discussion above, $\mathrm{B}-\mathrm{V}>0$ means the star is more *red* than Vega. $\mathrm{B}-\mathrm{V}<0$ means the star is more *blue* than Vega; it has relatively more energy in the blue. $\mathrm{U}-\mathrm{B}>0$ means also more energy at the longer wavelenghts, where 'longer' now means the blue wavelength band as compared to the ultraviolet, and $\mathrm{U}-\mathrm{B}<0$ means the star has relatively more energy in the ultraviolet than Vega.

Since in the colors we use V instead of m_{V} many astronomers now

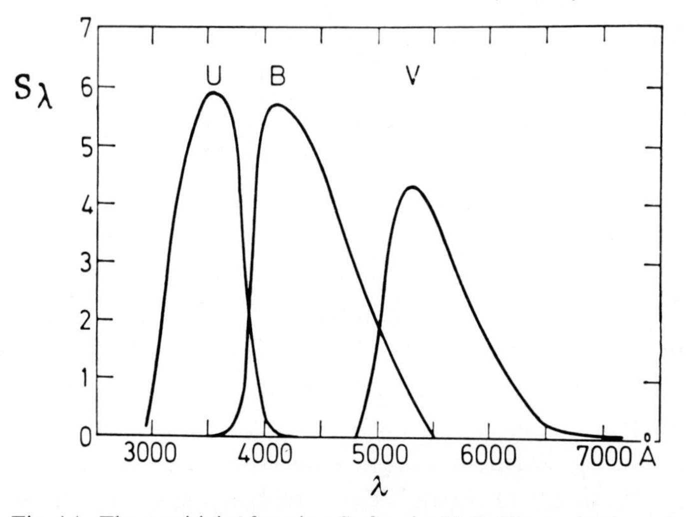

Fig. 4.1. The sensitivity function S_λ for the U, B, V magnitude scales, i.e., the relative intensities that would be measured through the U, B, V filters, for constant (wavelength independent) intensities I_λ, are shown. (From Unsöld, 1982.)

Table 4.1. *The relation between $B-V$, absolute magnitude M_v and $U-B$ colors for main sequence stars*

B − V	M_v Main sequence	U − B Main sequence
−0.30	−4.0	−1.08
−0.20	−1.6	−0.71
−0.10	−0.4	−0.32
0.0	+0.6	0.00
+0.10	+1.4	+0.10
+0.20	+2.1	+0.11
+0.30	+2.7	+0.07
+0.40	+3.3	+0.01
+0.50	+4.0	+0.03
+0.60	+4.6	+0.13
+0.70	+5.2	+0.26
+0.80	+5.8	+0.43
+0.90	+6.3	+0.63
+1.00	+6.7	+0.81
+1.10	+7.1	+0.96
+1.20	+7.5	+1.10
+1.30	+8.0	+1.22

abbreviate m_V by V. So V means the same as m_V.

In Table 4.1 we list the relation between the U − B and B − V colors for 'normal' (so-called main sequence) stars. These are, however not the colors as we measure them directly. Before we can say anything about the true colors of the stars we must correct the measured intensities for the absorption in the Earth's atmosphere.

4.3 Correction for the absorption of radiation in the Earth's atmosphere

When we measure the radiative energy arriving at the telescope we measure the radiation after it has passed through the Earth's atmosphere, which means after the Earth's atmosphere has absorbed a certain fraction of the light. This fraction is not always the same becasue we observe different stars at different positions in the sky, and even the same star is observed at different positions in the sky depending on the time of the year and the time of the night. Depending on the zenith distance z of the star the path length of the light beam of the star through the atmosphere is different, see Fig. 4.2, and that means that different amounts of light are absorbed during the passage through the atmosphere. If we want to know the amount of radiation arriving above the Earth's atmosphere – this is the only quantity which truly tells something about the stars – then we have to correct for the absorption in our atmosphere, also called the atmospheric extinction.

Duc to the extinction in the Earth's atmosphere, the light is reduced by a certain amount which is proportional to the intensity I_λ of the beam at wavelength λ. (The more photons there are passing through the atmosphere the larger is the chance that one of them will hit an atom and will be absorbed by the atom.) The chances of absorption are also larger if the path lengths through the atmosphere are larger. It is also large if the atoms are of a kind which want to absorb especially light of the wavelength λ considered. The properties of atoms concerning the absorption of light of a given wavelength λ are described by the absorption coefficient κ_λ per cm, which may be very strongly dependent on the wavelength λ. We then find that the intensity change $\mathrm{d}I_\lambda$ along the path element $\mathrm{d}s$ is

$$\mathrm{d}I_\lambda = -\kappa_\lambda \cdot I_\lambda \,\mathrm{d}s. \tag{4.5}$$

Dividing by I_λ and remembering that $\mathrm{d}I_\lambda / I_\lambda = \mathrm{d}(\ln I_\lambda)$ we find

$$\mathrm{d}(\ln I_\lambda) = -\kappa_\lambda \,\mathrm{d}s = -\mathrm{d}\tau_\lambda. \tag{4.6}$$

Here we have defined the so-called optical depth τ_λ by

$$\mathrm{d}\tau_\lambda = \kappa_\lambda \,\mathrm{d}s \quad \text{and} \quad \tau_\lambda(s_0) = \int_0^{s_0} \kappa_\lambda \,\mathrm{d}s. \tag{4.7}$$

Equation (4.6) can be integrated on both sides between 0 and s and yields

$$\Delta(\ln I_\lambda) = \ln I_\lambda(s) - \ln I_\lambda(0) = -\int_0^s \kappa_\lambda \,\mathrm{d}s = -\int_0^{\tau_\lambda(s)} \mathrm{d}\tau_\lambda = -\tau_\lambda(s), \tag{4.8}$$

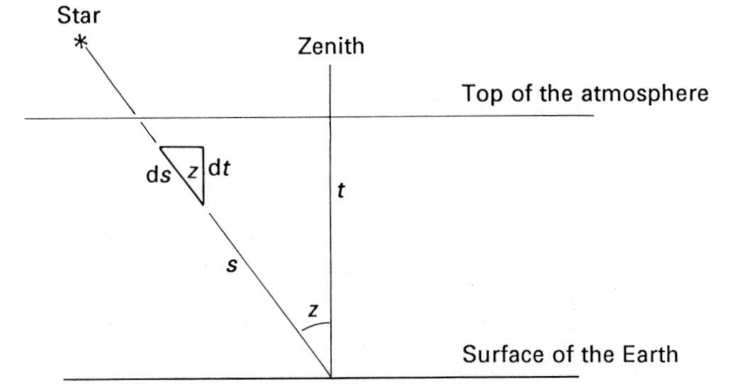

Fig. 4.2. Stellar light is entering the atmosphere at an angle z with respect to the zenith direction (exactly overhead). Light is absorbed along the path s. The longer s the more light is absorbed.

where τ_λ is the variable optical depth along the path s.

Taking the exponential on both sides gives

$$I_\lambda(s) = I_\lambda(0)\mathrm{e}^{-\tau_{\lambda s}(s)}. \tag{4.9}$$

The optical depth along the path of the light $\tau_{\lambda s}$ depends on the zenith distance z, as can be seen in Fig. 4.2. We see that $\cos z = t/s = \mathrm{d}t/\mathrm{d}s$, or

$$\mathrm{d}s = \frac{\mathrm{d}t}{\cos z} = \sec z\,\mathrm{d}t. \tag{4.10}$$

It then follows that

$$\tau_{\lambda s} = \int_0^s \kappa_\lambda\,\mathrm{d}s = \sec z \int_0^t \kappa_\lambda\,\mathrm{d}t = \sec z\,\tau_{\lambda t}, \tag{4.11}$$

where $\tau_{\lambda t}$ is the optical depth measured perpendicularly through the atmosphere. We can then write equation (4.9) in the form

$$I_\lambda(s, z) = I_\lambda(0)\mathrm{e}^{-\sec z \tau_{\lambda t}}, \tag{4.9a}$$

where $\tau_{\lambda t}$ is now independent of z.

$\tau_{\lambda t}$ is called the optical depth of the atmosphere at wavelength λ and is usually written at τ_λ.

In order derive from $I_\lambda(s, z)$ the intensity above the Earth's atmosphere we have to know τ_λ, which may depend strongly on the wavelength λ.

We can in principle determine τ_λ from two measurements of $I_\lambda(s, z)$ to give us two equations for the two unknowns $I_\lambda(0)$ and τ_λ. Let $I_{\lambda 1} = I_\lambda(z_1, s)$ be the intensity measured at zenith distance z_1 and $I_{\lambda 2}$ be the intensity measured at zenith distance z_2. Then, according to equation (4.9a)

$$\ln I_{\lambda 1} - \ln I_{\lambda 2} = -\tau_\lambda(\sec z_1 - \sec z_2)$$

and

$$\tau_\lambda = \frac{\ln I_{\lambda 1} - \ln I_{\lambda 2}}{(\sec z_2 - \sec z_1)} \tag{4.12}$$

As there are always measuring errors, it is safer to make many measurements and plot the result in a diagram, as shown in Fig. 4.3. The best fitting straight line gives the relation $\ln I_\lambda = \ln I_{\lambda 0} - \sec z \cdot \tau_\lambda$. (Here we have written $I_{\lambda 0}$ for $I_\lambda(0)$). The gradient of the line is determined by τ_λ, and $\ln I_{\lambda 0}$ can be read off on the $\ln I_\lambda$ axis from the extrapolation of the best fitting line to $\sec z = 0$. (Never mind that $\sec z = 0$ does not actually exist; this is just a convenient way to read off $\ln I_{\lambda 0}$).

Actually, Fig. 4.2 is a simplification of the situation, since the Earth's surface is not plane parallel and, in addition, the beam of light is bent due to refraction in the atmosphere. These effects give a relation $\tau_{\lambda s}/\tau_{\lambda t} \neq \sec z$. The actual ratio is called the air mass. For $\sec z < 2$ the difference is in the third decimal place and is in most cases negligible.

It is important to note that the above derivation applies only to each given wavelength λ with a given τ_λ. It cannot be applied to a broad wavelength band with varying κ_λ. Fig. 4.4. and 4.5 show the overall gross variations of

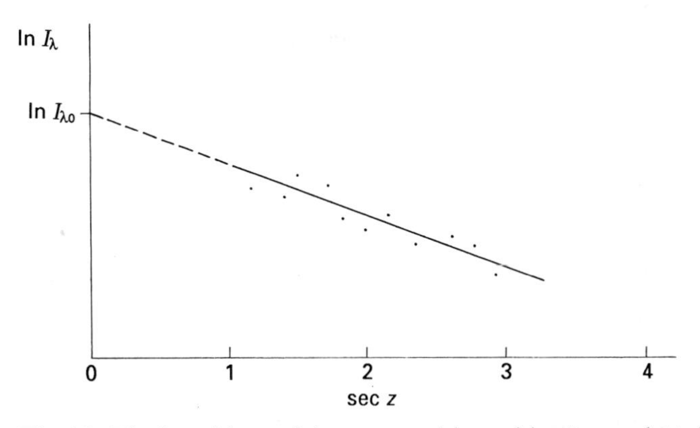

Fig. 4.3. The logarithms of the measured intensities I_λ are plotted as a function of sec z. The best fitting straight line through these points can be determined. The intersection of this line with the ln I_λ axis determines the logarithm of the intensity above the Earth's atmosphere, ln $I_{\lambda 0}$.

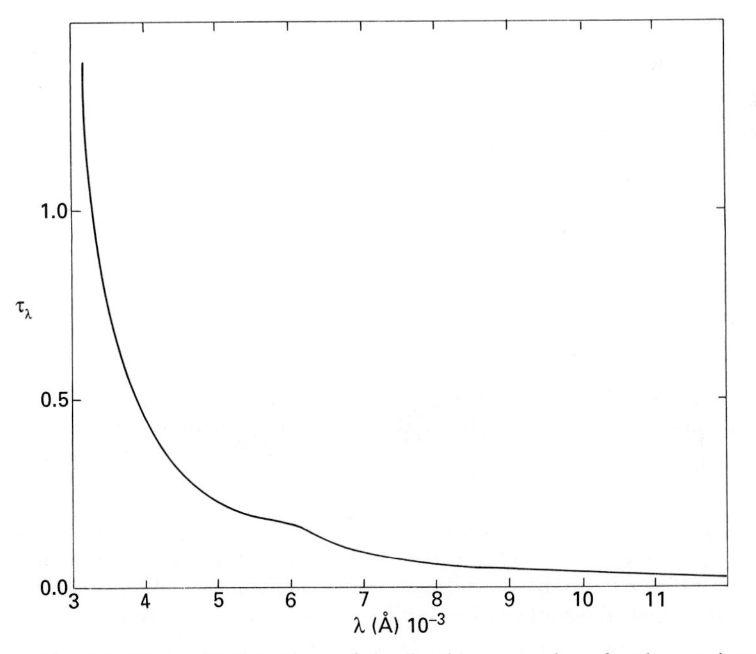

Fig. 4.4. The optical depth τ_λ of the Earth's atmosphere for the continuous absorption is shown as a function of wavelength λ, according to Allen (1968).

the absorption in the Earth's atmosphere with wavelength. Within the broad molecular bands seen in Fig. 4.5, which are due mainly to the absorption by the water molecules in the atmosphere, we actually have very many narrow lines. The actual variation of κ_λ with λ is much stronger than shown. If we use the method described above to determine the average τ_λ for such wavelength regions we may get very wrong values for $I_{\lambda 0}$.

According to (4.9a), the correction factor for the measured intensity is $e^{\sec z \cdot \tau_\lambda}$. Suppose we have measured star 1 at zenith distance z_1 and measured $I_{\lambda 0}(1) \cdot e^{-\sec z_1 \cdot \bar{\tau}_\lambda}$. For Star 2 we measured $I_\lambda(2) = I_{\lambda 0}(2) \cdot e^{-\sec z_2 \cdot \bar{\tau}_\lambda}$. We now determine

$$\frac{I_{\lambda 0}(1)}{I_{\lambda 0}(2)} = \frac{I_\lambda(1) \cdot e^{\sec z_1 \cdot \bar{\tau}_{\lambda 1}}}{I_\lambda(2) \cdot e^{\sec z_2 \cdot \bar{\tau}_{\lambda 2}}} = \frac{I_\lambda(1)}{I_\lambda(2)} \cdot e^{(\sec z_1 - \sec z_2) \cdot \bar{\tau}_\lambda} \tag{4.13}$$

if $\bar{\tau}_{\lambda 1} = \bar{\tau}_{\lambda 2} = \bar{\tau}_\lambda$. If we have determined a wrong $\bar{\tau}_\lambda$ the error is minimized if z_1 is chosen as close to z_2 as possible. It is zero if $z_1 = z_2$. The error is also smaller if we choose small z_1 and z_2 because then also $\sec z_1$ and $\sec z_2$ will be small and the difference remains small, even if $z_1 \neq z_2$.

If we measure stars with very different energy distributions then $\bar{\tau}_{\lambda 1} \neq \bar{\tau}_{\lambda 2}$. In this case the error does not cancel out, even if $z_1 = z_2$, but the error is still minimized for $z_1 = z_2$ and small $\sec z$.

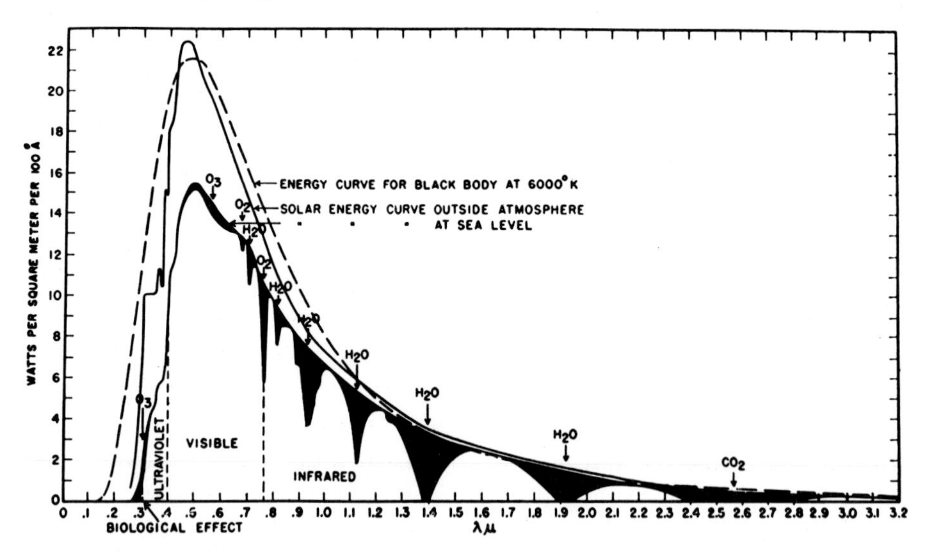

Fig. 4.5. The measured energy distribution of the sun, which means the amount of radiation which we receive from the sun here on the surface of the Earth, is plotted as a function of wavelength of the radiation. The black areas indicate the amount of radiation taken out by the molecular bands of water vapor and of oxygen. Also shown is the energy distribution as measured above the Earth's atmosphere. For comparison we also show the energy distribution of a black body with a temperature of 6000 K, which matches the solar energy distribution rather well. (From Pettit 1951.)

4.4 The black body

4.4.1 *Definition of a black body*

If different stars have different colors, the colors must tell us something about the nature of the stars and their differences. In order to find out about the possible meaning, we have to compare the colors of the stars with colors of known objects on Earth. We do not learn anything if we compare the stars with just *any* colored object on Earth, we need some sort of *ideal source of light* whose properties depend only on one parameter. Such a source is the so-called 'black body', whose radiation depends only on the temperature.

What is a black body? We would call an object black if it looks black in bright daylight, this means that the object does not send out any light rays that enter our eyes. The light that our eyes receive from objects which do not look black in bright daylight is usually not light which these objects emit of their own but is solar light that is reflected by these objects. Solar light appears yellow to us. Most objects on Earth do not look yellow even though we see only solar light reflected by them. The reason for the difference in color is the fact that any object absorbs part of the solar light, and only the part which is not absorbed can be reflected. If mainly blue wavelengths were absorbed, then the object will appear more red than the sun, if mainly red wavelengths were absorbed then the object will appear more blue. If all wavelengths are strongly absorbed, then no light is left to be reflected into our eyes and the object will appear black. We therefore define a *black body* as *an object that absorbs all light* which falls on it. This does not mean that a black body always has to look black. A black body may generate radiation by itself, then it may actually be quite bright even though it absorbs all the light reaching it from other light sources. Think about a hot plate on an electric stove. If it is not turned on, it looks almost black because it absorbs nearly all the light which falls on it, but when you turn it on, it becomes hot and now produces its own light. It then shines brightly even if you turn off all the outside lights. The hot plate is still nearly a black body because it still absorbs nearly all the light falling on it but it certainly does not look black any more. When you turn the hot plate on 'LOW' you will see it glow in a dark red color, when you turn it on higher you will find that it becomes brighter and more yellowish; if it becomes still hotter, it will shine still brighter and more bluish. In this example, we see already that the color of such a near black body tells us something about the temperature of the object.

How can we realize an ideal black body, which means an object that absorbs all the light falling on it? The best way to do this is to make a small hole in a box, see Fig. 4.6. A light beam falling into the hole will hit a wall in the box and will be reflected, part of it may be absorbed by the wall. The reflected part will hit another wall and will be partly absorbed and partly reflected again and so on. Chances that even a small fraction of the light might finally find its way back out through the hole are extremely small and will be smaller the smaller the hole is. The hole is indeed a black body. Actually the windows in a building are approximations to this kind of a black body; when you look at the windows from the outside, they appear black, except if you put a light source inside this black body, such as a burning electric bulb, then the black body radiates on its own merits.

If we insulate such a black body very well from its surroundings and leave it to itself for a long time, then the inside of the box, shown in Fig. 4.6, achieves an equilibrium, which means that nothing in the box will change anymore in time or in space. We say the box has reached thermodynamic equilibrium. In physics we only call a box with a hole a black body if it has achieved thermodynamic equilibrium. This is included in the definition of a 'black body.'

4.4.2 *Radiation of a black body*

For a black body we can measure the radiation that comes out of the hole, it is radiation emitted by the black body itself, either by gas in it or by the inside walls of the box. For the radiation of such a black body in thermodynamic equilibrium we can measure that the radiation coming out of the hole is independent of the material of the box and independent of the gas in the box. No matter which kind of box we take, *after thermodynamic equilibrium is achieved* the radiation is always the same if the temperature of the box is the same, but the radiation changes if the temperature of the black

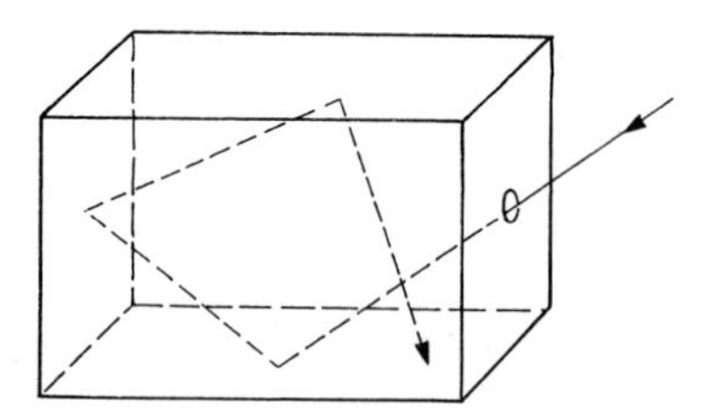

Fig. 4.6. A black body can best be realized by a small hole in a box. Light falling into the hole in the box will be reflected back and forth and will not find its way out of the hole before being absorbed at the walls. For a black body, as defined in physics, we have to insulate the box very well from its surroundings and leave it to itself for a long time so that thermodynamic equilibrium can be achieved in the box.

body changes. The change is similar to the changes seen for the hot plate with increasing temperature. For higher temperatures, more radiation is emitted and the radiation looks more blue. From such measurements of the radiation of a black body we find that the total radiative energy E emitted per cm^2 and per s in all directions by a black body of temperature T increases with the fourth power of the temperature, i.e.

$$E = \sigma T^4 = \pi F. \tag{4.14}$$

where σ is the so-called Stefan–Boltzmann constant and is $\sigma = 5.67 \times 10^{-5}\,\mathrm{erg\,cm^{-2}\,s^{-1}\,K^{-4}}$, F is called the flux. Equation (4.14) is the *Stefan–Boltzmann law.*

4.4.3 *Spectral energy distribution*

If we measure the amount of radiation per wavelength interval of 1 cm coming out of the hole per s perpendicular to the area of the hole and into a cone with a solid angle $\Delta\omega = 1$, and if we divide this amount by the area of the hole, then this energy is called the intensity I_λ (see Fig. 4.7)

In Fig. 4.8 we show the wavelength dependence of the intensities emitted by black bodies of different temperatures.

Of course, we cannot make black bodies with temperatures of 6000 K in the laboratory because any box of any kind of material would melt at such temperatures, but we can derive the law for the intensity distribution from measurements for lower temperatures and then use this law to calculate what the energy distribution for higher temperatures will look like. You will recall that it was the observed law for the radiation of a black body which led Max Planck to the discovery of the light quanta.

For a black body the intensity $I_\lambda = B_\lambda$ is given by

$$I_\lambda = \frac{2hc^2}{\lambda^5} \frac{1}{e^{hc/\lambda kT} - 1} = B_\lambda = \text{Planck function.} \tag{4.15}$$

Here, c = velocity of light, k = Boltzmann constant, and h = Planck constant.

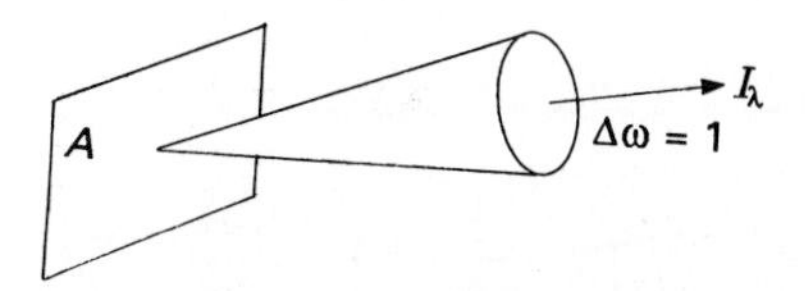

Fig. 4.7. The energy emitted from a surface area A of 1 cm^2 per s per wavelength band $\Delta\lambda = 1$, perpendicular to A and into a cone with opening $\Delta\omega = 1$, is called the intensity I_λ. For a black body the intensity $I_\lambda = B_\lambda =$ Planck function, given by (4.15).

We certainly do not measure the intensity with a wavelength band of 1 cm because the intensity would vary by very large factors over such a wide band. What we mean is that we measure the intensity in a small wavelength band $\Delta\lambda$ and then divide the measured energy by the wavelength interval $\Delta\lambda$. The amount of energy received will of course increase if the bandwidth is increased.

Keeping in mind the energy distributions of black bodies and the increasingly more blue color for higher temperatures we may suspect that the blue stars have higher temperatures than the red stars.

4.4.4 *Wien's displacement law*

The intensity distribution of a black body as given by (4.15) has a very important property. As we see in Fig. 4.8, the maximum shifts to shorter wavelengths for higher temperatures. From (4.15) it can be easily derived that the wavelength λ_{max}, for which the maximum intensity is measured. (i.e. the wavelength λ_{max} for which $dI_\lambda/d\lambda = 0$) is inversely proportional to the temperature T, or in other words,

$$\lambda_{max} \cdot T = \text{const.} = 0.2897\,\text{cm} \cdot \text{degree} \tag{4.16}$$

if the wavelength λ_{max} is measured in cm.

Equation (4.16) offers an easy way to determine the temperature of a black body without having to measure the whole energy distribution, we only have

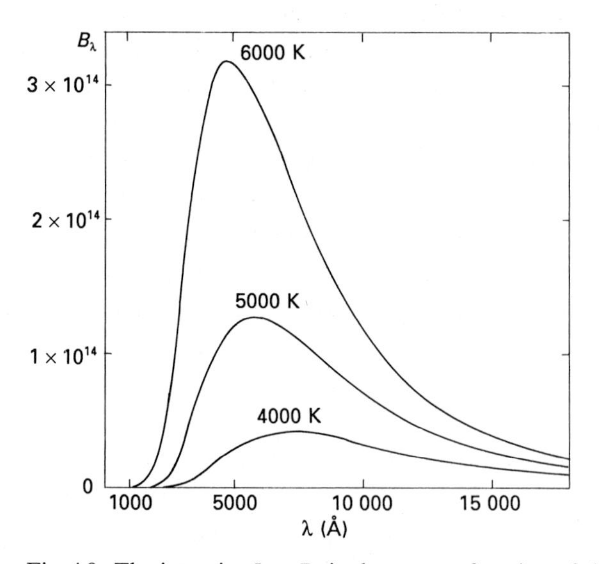

Fig. 4.8. The intensity $I_\lambda = B_\lambda$ is shown as a function of the wavelength λ for black bodies with different temperatures T, as indicated on the curves.

to determine the wavelength at which the energy distribution has its maximum. If the stars were to radiate like black bodies we could easily determine their temperatures in this way.

4.4.5 *The colors of black bodies*

Because black bodies become more blue with increasing temperature we suspect that the colors of the stars tell us something about the temperatures of the stars. We expect the blue stars to have higher temperatures than the red stars because for black bodies the radiative energy shifts to shorter wavelengths for higher temperatures. In order to check how well the colors of the stars agree with those of black bodies we have to determine the B – V and the U – B colors for black bodies with different temperatures. This means we have to determine the U, B, V magnitudes of the black bodies which in turn requires us to compare the brightness of the black bodies in the U, B and V bands with the brightness of Vega. Unfortunately, this is not very easy. In order to have the same transmission functions for the star and the black body we have to observe the star and the black body with the same instrument, namely with the telescope. The telescope gives only sharp images of light sources which are very far away. We therefore have to put the black body at a large distance which makes it rather faint.

Even more of a problem is the correction for extinction in the Earth's atmosphere. The beam of light from Vega passes through the whole atmosphere, all layers of the atmosphere with their different absorption coefficients contribute to the extinction. For the black body, which we may put at the next mountain top, only the lower layers of the atmosphere contribute to the extinction. We would have to know the height dependence of the atmospheric absorption coefficient in order to make the correct extinction correction for the black body. While from the physical point of view it would be more logical to compare the stellar brightnesses directly with those of black bodies to begin with, this causes so many difficulties that it is intrinsically much more accurate to compare stars with stars. The final comparison with black bodies does not then influence the intrinsic comparison of stars among themselves.

When we do the best we can for the determination of the colors of black bodies, we find the B – V and U – B colors for black bodies as shown in the two color diagram, Fig. 4.9. Table 4.2 shows the actual numbers. Every point on the black body curve in Fig. 4.9 corresponds to a black body with a given temperature, as indicated for a few values of the temperature. Remember stars with positive values of B – V are more red than Vega. B – V therefore increases for lower temperature. The same holds for U – B.

In the same diagram we have also entered values measured for nearby stars. They also seem to follow one curve, which is, however, different from the one for black bodies, but still all stars with a given color B – V have nearly the same U – B color. Both colors are determined by one parameter, which we still suspect to be the temperature. Before we can, however, establish a relation between the colors of a star and its temperature we have to understand the difference between the stellar energy distributions and those of black bodies, and especially why the stars have so much less energy in the U band than the black bodies. This we will discuss in volume 2, when we will discuss the radiation and the structure of stellar atmospheres.

4.5 The solar radiation

The brightest star is of course our sun. We can measure the amount of radiative energy received from the sun and correct it for the absorption in

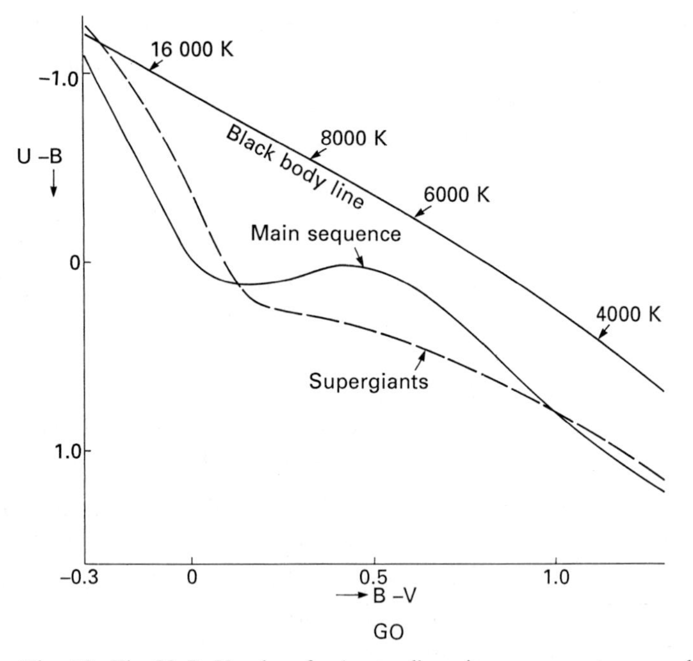

Fig. 4.9. The U, B, V colors for 'normal', main sequence stars are shown in a two-color diagram. For increasing B – V colors the U – B colors generally also increase, except in the range $0.1 < B - V < 0.5$. Also shown are the colors for supergaints, see chapter 10.2, for which the U – B colors always increase with increasing B – V. We have also plotted the U – B versus B – V colors for black bodies. For a given B – V the stars have less radiation in the ultraviolet than black bodies.

Table 4.2 *The B − V and U − B colors of black bodies with different temperatures according to Lamla, 1982*

B − V	U − B (±0.03) Black body
−0.30	−1.22
−0.20	−1.11
−0.10	−1.00
0.00	−0.88
+0.10	−0.76
+0.20	−0.66
+0.30	−0.54
+0.40	−0.44
+0.50	−0.33
+0.60	−0.22
+0.70	−0.10
+0.80	+0.02
+0.90	+0.14
+1.00	+0.25
+1.10	+0.37
+1.20	+0.49
+1.30	+0.61

the Earth's atmosphere in exactly the way described in the previous section. We can now also measure the total solar radiation directly, outside the Earth's atmosphere from satellites. The satellites called 'Space Lab' and the 'Solar Maximum Mission' (the first satellite to be repaired while in orbit) have made very careful measurements of the solar radiation. What is found is an energy distribution, also shown in Fig. 4.5. The continuum energy distribution, which means the energy distribution between the spectral lines, looks very similar to that of a black body at 6000 K. The energy has a maximum around 5000 Å, the wavelengths at which our eyes are most sensitive, luckily enough also in the wavelength region in which the Earth's atmosphere is rather transparent.

How much energy do we actually get from the sun altogether? The total radiation received from the sun per s per cm^2, perpendicular to the beam of light above the Earth's atmosphere, is called the solar constant S. It is measured to be $S = 1.368 \times 10^6\,\mathrm{erg\,cm^{-2}\,s^{-1}}$.

Since there is so much talk about solar energy as the energy source for our industrialized life, let us stop for a moment and realize how much that actually is.

Since 1 square foot has about 10^3 cm^2 and since 10^7 erg sec^{-1} is 1 watt we can also write $S \approx 0.14 \times 10^3$ watt per ft^2 or $S \approx 0.14$ kW per ft^2.

If your room has a roof area of about 200 ft^2 and the sun shines on average 5 hours per day, you would collect on your roof energy of the order of 140 kW hours each day. This is not really correct since the sun is not shining perpendicularly on your roof and there is some absorption in the atmosphere, but it gives the correct order of magnitude, it may be a factor of 2 or 3 less. This would still be plenty to cover your personal needs. Certainly for a large industrial plant the sun shining on their roof would not be sufficient, but our main problem still is that we do not yet know how to convert the solar ergs per s into kW of electrical energy efficiently. Currently we lose a factor of between 10 and 100 in the conversion.

4.6 The absolute magnitudes of stars

The apparent magnitudes do not really tell us anything about the intrinsic brightness of the stars which to the astrophysicist is much more important than the apparent brightness. A star may be intrinsically very bright and just appears very faint because it is so distant. If we can determine the distance we can correct for this effect. We can then in our minds put all the stars at the same distance and compare their intrinsic brightnesses. The magnitudes which the stars would have if they were at a distance of *10 pc* are called the *absolute magnitudes*. They are designated by a capital M. Again we have absolute magnitudes M_V, M_B, M_U for the different wavelength bands.

Vega is at a distance of 8.4 pc, i.e., is closer than 10 pc. If Vega was at a distance of 10 pc it would be fainter. Its absolute magnitude is therefore larger than its apparent magnitude: $M_V(\text{Vega}) > 0$. It comes out to be $M_V(\text{Vega}) = 0.5$. We have to calculate the magnitude which Vega would have if it were at the distance of 10 pc and compare this with the magnitude which it actually has, namely $m_v = 0$.

In order to determine the absolute magnitude we have to know the distance d of the star

$$\frac{I(10\,\text{pc})}{I(d)} = \left(\frac{d}{10\,\text{pc}}\right)^2 \quad \text{because} \quad I \propto \frac{1}{d^2} \tag{4.17}$$

$$M_V = m_v(10\,\text{pc})$$

$$M_V - m_v = -2.5[\log I(10\,\text{pc}) - \log I(d)]$$

$$= -2.5[2\log d - 2\log 10\,\text{pc}]$$

$$= -5[\log d[\text{pc}] - 1]$$

$$= -5\log d[\text{pc}] + 5 \tag{4.18}$$

$m_v - M_V$ is called the distance modulus of the star because it is determined by the distance of the star.

For stars with parallax angles $\pi > 0''.05$ we can measure π fairly accurately. We can therefore determine their distances from their parallaxes. For these stars we can also determine their absolute magnitudes M_U, M_B, M_V.

What are the stellar colors in absolute magnitudes?

From (4.18) we see

$$M_V = m_v - 5\cdot\log d[\text{pc}] + 5 \tag{4.19}$$

$$M_B = m_B - 5\cdot\log d[\text{pc}] + 5 \tag{4.20}$$

and

$$M_B - M_V = \text{B} - \text{V} = m_B - m_V. \tag{4.21}$$

This means we can always determine the colors from the apparent magnitudes. The absolute colors are the same. See, however, Chapter 19 for interstellar reddening.

5

Color magnitude diagrams

5.1 Color magnitude diagrams of nearby stars

For nearby stars, say within 20 pc, we can determine the distances from trigonometric parallaxes. From the apparent magnitudes and the distances we can calculate the absolute magnitudes, i.e., the magnitudes which the stars would have if they were at a distance of 10 pc. This means that for absolute magnitudes we compare the brightness the star would have if it were at a distance of 10 pc with the actual brightness of Vega at its actual distance, i.e. with its apparent brightness. It turns out to be quite instructive to plot the absolute magnitudes of the stars as a function of their B – V colors. In Fig. 5.1 we do this for the nearby stars. While we might have expected that stars with a given color could have quite different absolute magnitudes, it turns out that this is generally not the case. Most of the stars with a given B – V color have the same absolute magnitude. Most of the stars fall along one line in the color magnitude diagram. This line is called the *main sequence*. The intrinsic brightnesses and the colors of these stars are obviously determined by just *one* parameter, since they follow a one-dimensional sequence. It turns out, as we shall see in Volume 3, that this one parameter is the mass of the star.

While most of the stars follow the main sequence we see a few which are outsiders. For a given value of the B – V color most of these stars have a larger intrinsic brightness than the main sequence stars, which means a smaller absolute magnitude. Because brighter stars have smaller magnitudes the smallest magnitudes are always plotted at the top in the color magnitude diagrams. On the other hand, the B – V colors increase towards the right in the color magnitude diagrams. If, as we suspect, and as is actually the case, the B – V colors increase for lower temperatures because the black bodies become more red for lower temperatures, then the temperatures decrease toward the right-hand side of the diagram.

Let us compare two stars with a given B – V, which means, as we suspect, with nearly the same temperature, but with different brightnesses. For a given temperature the radiation of a black body per cm^2 increases with T^4 as we saw in the previous chapter. For a given temperature the total radiation of a black body per cm^2 is always the same. If the B − V colors of the stars are a measure of the temperatures then stars of a given B − V should radiate approximately the same amount of energy per cm^2. If the stars were to radiate like black bodies their total radiation, called their luminosity L should be

$$L = 4\pi R^2 \sigma T^4, \tag{5.1}$$

where R is the radius of the star and therefore $4\pi R^2$ is the surface area of the star. If two stars with a given B − V radiate vastly different amounts of energy as shown by their different absolute magnitudes, then the brighter star must have the larger surface area, which means it must be larger. The

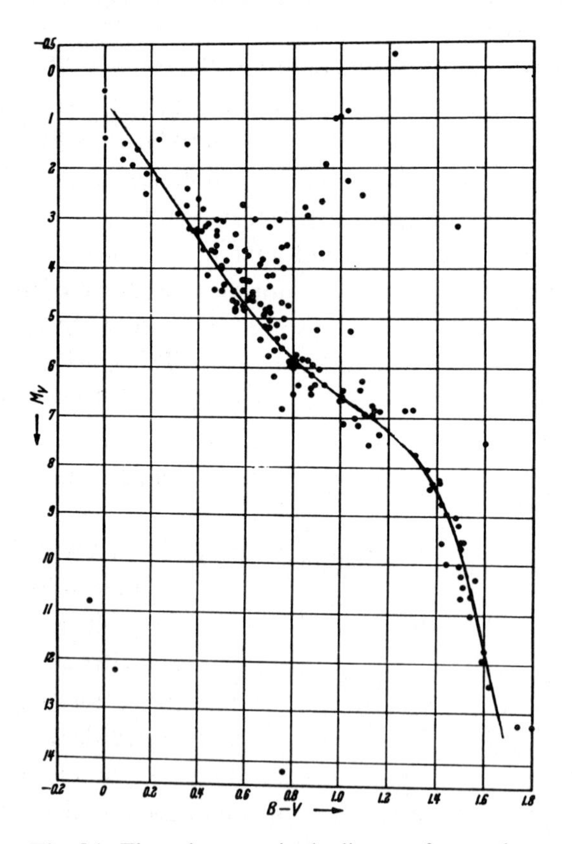

Fig. 5.1. The color magnitude diagram for nearby stars is shown according to Johnson and Morgan (1953). For better comparison with the following color magnitude diagrams we have added an eye-fitted average curve for the main sequence stars.

stars which are brighter than the main sequence stars are therefore called giants. In contradistinction to the giants the main sequence stars are called dwarfs. In astronomy, we have either giants or dwarfs but nothing in between (except perhaps subgiants; see Section 5.3). When we now look at the stars which are fainter than main sequence stars, we know that these stars must be smaller than the main sequence stars if they have the same B – V but less radiation. Since the name 'dwarf' has been used already for the main sequence stars we are in trouble, we do not know what to call them. Most of these faint small stars have rather bluish or white colors. They are therefore called white dwarfs. Not all of them are actually white, some of them are rather blue and a few are also red. So now we have blue (or hot) 'white dwarfs' and we have fairly red (or cool) 'white dwarfs', which sounds confusing. But, as always in astronomy, there is a long tradition and it would be even more confusing to change the names now.

5.2 Color magnitude diagrams for open clusters

Since we see this one-dimensional main sequence for the nearby stars we might wonder whether this is also true for the stars at larger distances. But how can we determine their absolute magnitudes? Fortunately, we do not have to know the distances of the stars which we plot in a color magnitude diagram as long as we know that they are all at the *same* distance. We could have plotted the same kind of color magnitude diagram for nearby stars, putting them in our minds all at a distance of 20 pc, then they would all have been fainter by a factor of 4, which means their magnitudes would all be larger by $\Delta m = 1.5$. The whole diagram would be shifted down by 1.5 magnitudes but otherwise remain the same. If we have a group of stars that are at the same distance, no matter what that distance is, we can still compare their intrinsic magnitudes. The unknown distance only introduces an upward or downward shift of all the magnitudes dependent on the distance, but this shift is the same for all the stars at the same distance. There are groups of stars visible in the sky which we can see are at the same distance, these are the star clusters. The best known one which is easily seen by the naked eye are the Pleiades. A photograph of this cluster is seen in Fig. 5.2(*a*). There are so many stars close together that it cannot be an accident. Most of these stars must actually belong together and form a star cluster. They must then be all at the same distance. The differences of the apparent magnitudes of these stars must then also be the differences of their absolute magnitudes. In Fig. 5.3 we show the color magnitude diagram with the absolute magnitudes for the Pleiades. Most of the stars again lie along

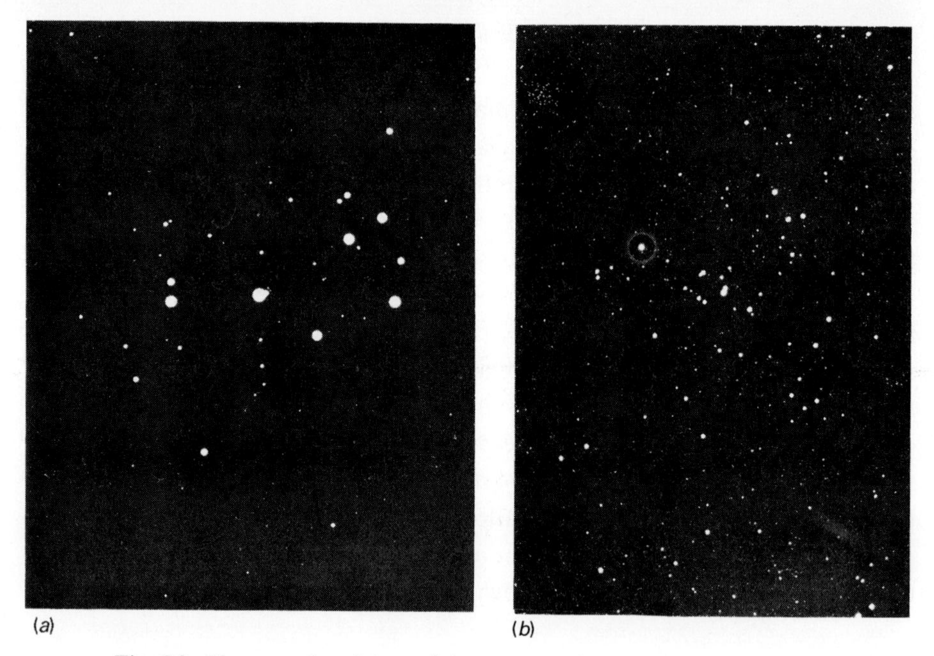

Fig. 5.2. Photographs of the well-known star clusters, the Pleiades (a), and the Hyades (b), in the constellation of Taurus. (From Burnham 1978.)

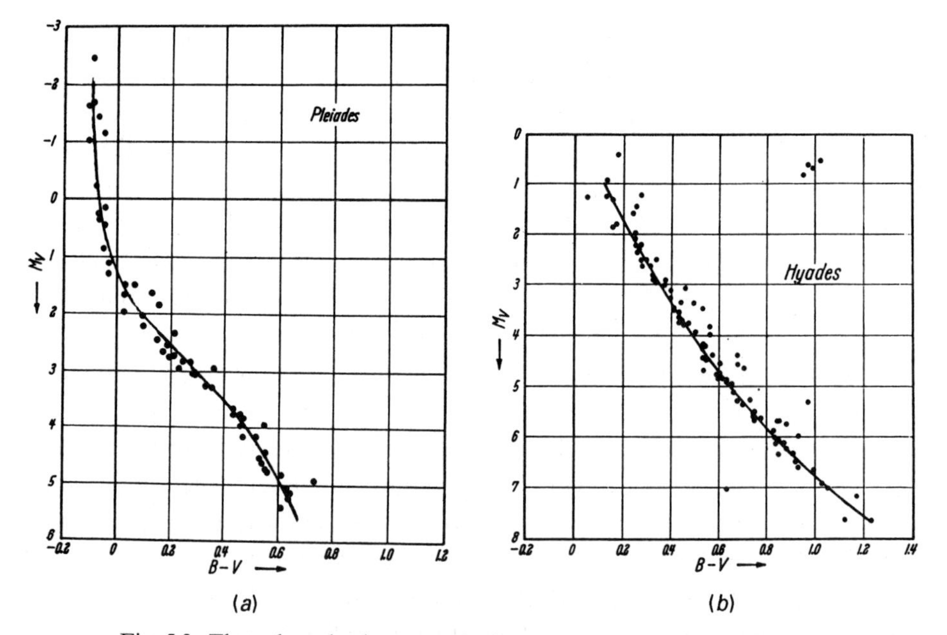

Fig. 5.3. The color, absolute magnitude diagrams for the Pleiades (a) and Hyades (b) star clusters are shown according to Arp (1958). A distance modulus $m_V - M_V = 5.3$ was used for the Pleiades. For the Hyades individual distance moduli were used, averaging $m_V - M_V = 3.08$. Distance moduli larger by 0.2 magnitudes would be considered more appropriate now.

For better comparison of the different main sequences we have added some eye-fitted average curves.

one sequence, probably the same main sequence which we see in Fig. 5.1, but the sequence for the Pleiades has to be shifted upwards to smaller magnitudes by $m_V - M_V \approx 5.3$. In Fig. 5.2(b) we reproduce a photograph of another well-known cluster, the Hyades, and in Fig. 5.3(*b*) we show the color magnitude diagram for this cluster. The difference between the m_V for a given B – V in the Hyades and the M_V for the nearby stars is $m_V - M_V = 3.3$.

There is, however, a distinction between the different color magnitude diagrams with respect to the bluest stars seen. In the Pleiades we can see stars which are much more blue than for instance in the Hyades. If we shift the lower parts of the main sequences on top of each other, as seen in Fig. 5.4, then the blue part of the main sequence for the Pleiades extends to smaller magnitudes or to intrinsically brighter stars.

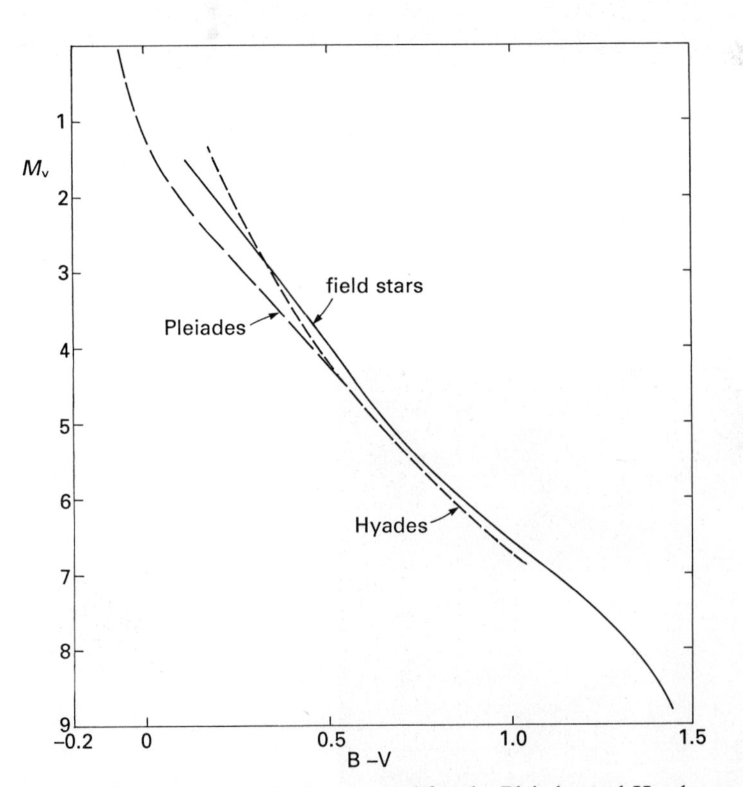

Fig. 5.4. For the nearby fields stars and for the Pleiades and Hyades we have plotted the eye-fitted average main sequences from Figs. 5.1 and 5.3 together in one color magnitude diagram. A better fit of the three sequences is obtained if the distance moduli of the Hyades and Pleiades are increased by 0.2 magnitudes, as compared to the distance moduli adopted for Figs. 5.1 and 5.3.

At the high luminosity end the main sequences diverge because the average ages of the field stars, the Hyades and the Pleiades are different. For further discussion see Volume 3.

If we had superimposed color magnitude diagrams of a large number of open clusters, we would have detected another group of stars which are even brighter than the giants. These very bright and therefore very large stars are called supergiants.

5.3 Color magnitude diagrams for globular clusters

So far we have shown only clusters with rather bright, blue stars. We can easily distinguish the different stars in these clusters. They are called open clusters. There are also clusters which look very different, these are the so-called globular clusters. In Fig. 5.5 we reproduce photographs of two of these globular clusters called M92 and M3. (They have the numbers 92 and 3 in the Messier Catologue of nebulous objects.) The globular clusters contain 10 000–1000 000 stars. In the centre they cannot all be resolved. Only in the outer parts of the clusters can we resolve single stars. When the colors and magnitudes of these stars were measured it came as a big surprise that the color magnitude diagrams for these clusters looked very different from the ones for the open clusters. In Fig. 5.6 we show the color magnitude diagrams for the globular clusters M92 and M3. You can hardly find a sequence which looks similar to the main sequence. The main branches seen are two nearly horizontal sequences of different magnitude and a nearly

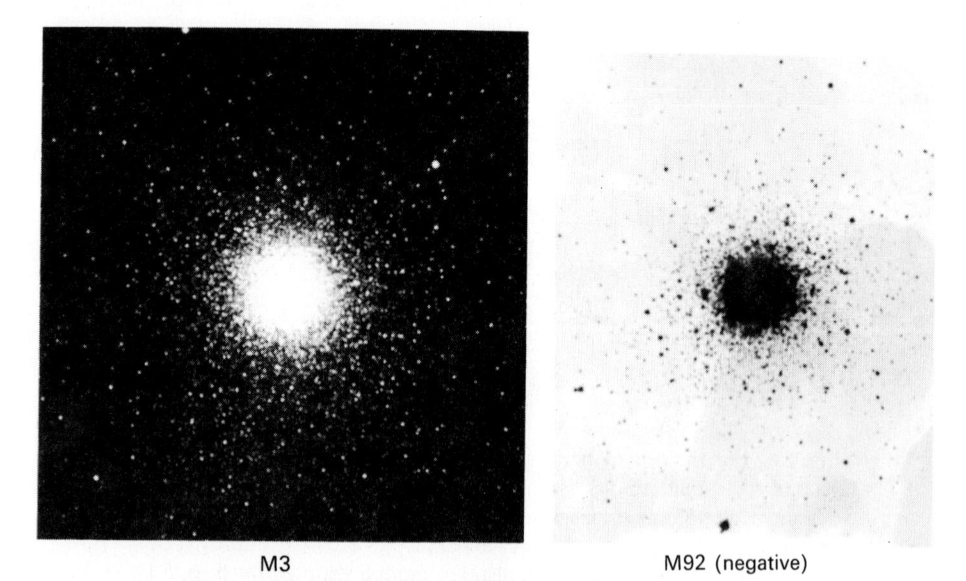

Fig. 5.5. Photographs of the two globular clusters M3 and M92 are shown. From Sandage and Walter (1966) and from Clayton (1983). Photographs from Mt. Wilson and Palomar Observatories.

vertical branch of red stars. On the blue side of the diagram we see, however, a short stub of a sequence which has the same inclination as the main sequence for the open clusters, but no bright blue stars are seen in these clusters. The color magnitude diagrams for other globular clusters all look

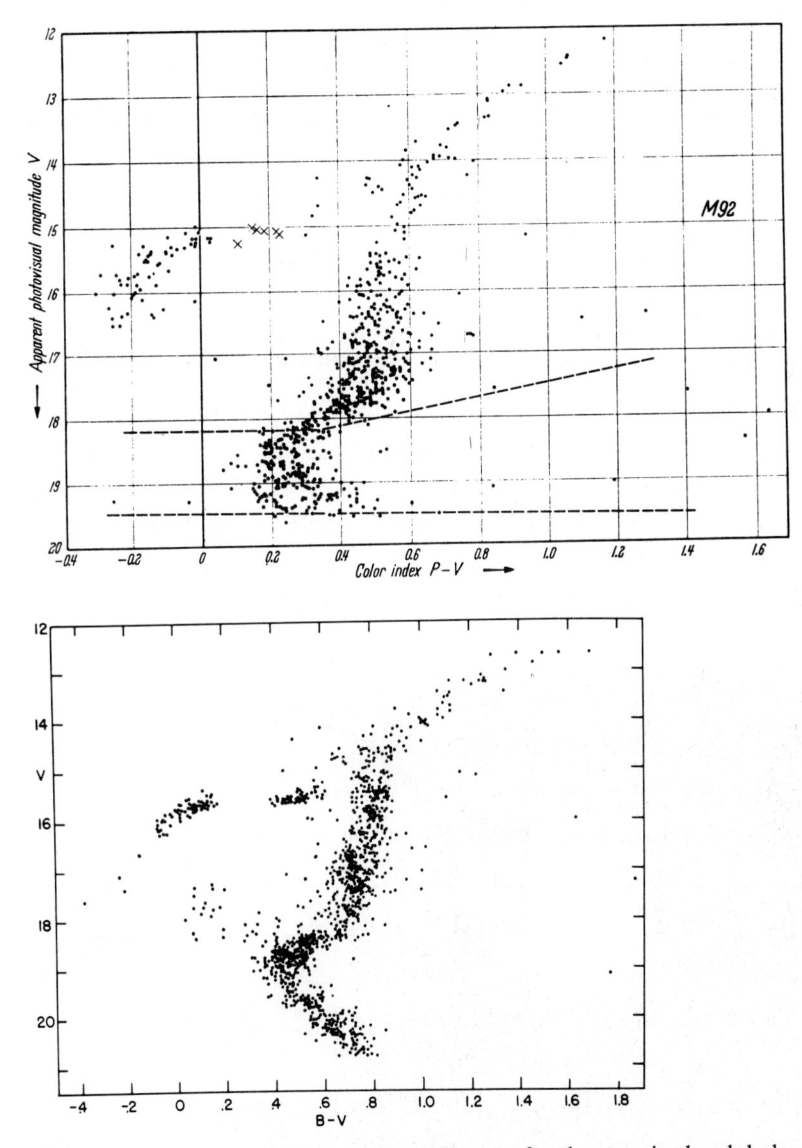

Fig. 5.6. The color apparent magnitude diagram for the stars in the globular cluster M92 is shown (top), according to Arp (1958). For the magnitude range between the dashed lines a smaller field of the cluster was used than for the brighter stars. *P* stands for photographic.

In the lower diagram the color apparent magnitude diagram for the stars in the globular cluster M3 is shown, according to Arp (1958).

very similar to these except that the upper horizontal sequence always looks different and sometimes consists only of a blue sequence or a short stub on the red side. The upper horizontal sequence is called the horizontal branch (HB). The lower horizontal sequence is called the subgiant branch and the nearly vertical sequence is called the red giant branch. The short stub of a sequence of faint stars in the diagram is indeed what is left over of the main sequence. In Fig. 5.7 we show a schematic color magnitude diagram of a globular cluster, in which the names for the different branches are given.

It is the main topic of Volume 3 to explain and understand the differences between the color magnitude diagrams for open and globular clusters. It is the big success of stellar evolution theory that we are actually able to understand most of the details of these diagrams.

5.4 Photometric parallaxes for star clusters

In Section 5.2 we have seen that the main sequences in the different color magnitude diagrams of open clusters can be shifted on top of each other by a vertical shift, i.e., a shift by Δm_V only.* This is to be expected if indeed all the main sequences agree but the clusters are at different distances. The necessary vertical shift $\Delta m_V = m_V - M_v$ is then actually the distance modulus for each cluster. This distance modulus determines the distance

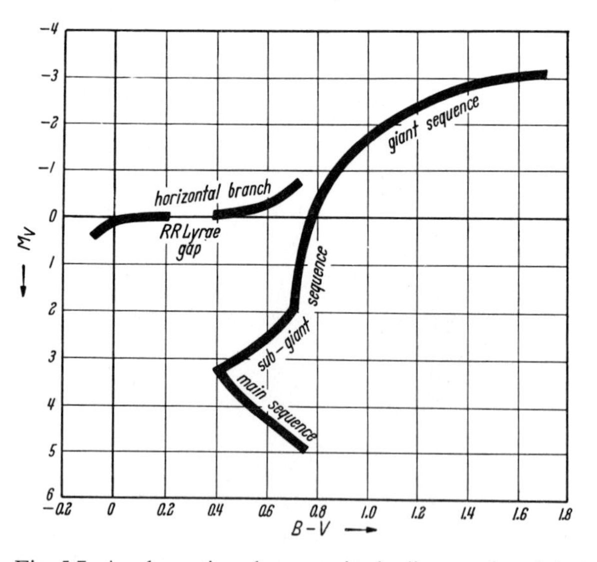

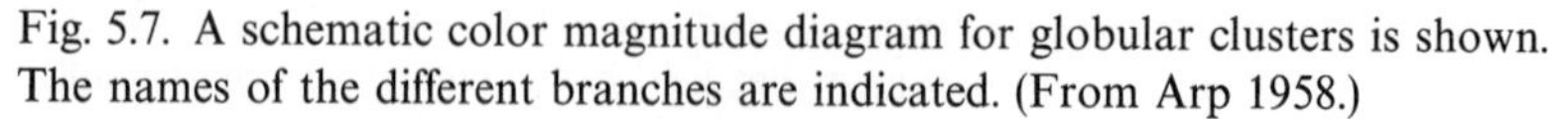

Fig. 5.7. A schematic color magnitude diagram for globular clusters is shown. The names of the different branches are indicated. (From Arp 1958.)

* Except for a color correction which compensates for reddening by interstellar absorption (see Chapter 19).

of the cluster, which can be derived from (4.21). For the Hyades with $m_V - M_V = 3.27$, we derive a distance of 45 pc, for the Pleiades with $m_v - M_V = 5.5$ we find a distance of 130 pc. By fitting the stub of the main sequence for globular clusters with the main sequence for nearby stars we can also determine the distances to the globular clusters if we can still see the faint stars of the left over piece of the main sequence. For very distant clusters, these stars are too faint to be observable except with the largest telescopes and modern receivers. These new receivers, the CCDs (Charge Coupled Devices) now permit us to observe the main sequences of globular clusters to much fainter magnitudes than shown in Fig. 5.6. They look indeed very similar to what we expected.

5.5 Photometric parallaxes for single stars

For main sequence stars we can also use this method to determine the distance to a given star. Since for main sequence stars the absolute magnitude M_V is a unique function of the B − V color, which we can read off from Fig. 5.1, we only need to measure the B − V color and the apparent magnitude m_V of the stars. For the measured B − V color we read off the M_V from Fig. 5.1.† The values of M_V for main sequence stars for different values of B − V are also given in Table 4.1. $m_V - M_V$ is the distance modulus from which the distance can be obtained again according to (4.18). These parallaxes are called photometric parallaxes.

Unfortunately, there are severe problems with this method of determining parallaxes for single stars;

1. The star could be a giant or supergiant or perhaps a white dwarf. We will see in Chapter 10 how we can distinguish these different types of stars by their spectra.

2. The color of the star could be altered by interstellar absorption as we will discuss in Chapter 19, where we will also see how we can determine this color change.

† Except for a color correction which compensates for reddening by interstellar absorption (see Chapter 19).

6

The luminosities of the stars

6.1 Fluxes, luminosities and angular radii

What is of interest to the astrophysicist even more than the amount of energy which is emitted by the star in a given wavelength band, for instance in the visual, is the total amount of energy emitted by a star. The radiation of the stars is emitted in all directions. We can therefore say that the same amount of energy goes through every cm^2 of a sphere of radius d around the star, see Fig. 6.1. If we are at a distance d from the star, then we can measure the amount of energy arriving here above the Earth's atmosphere per cm^2. We denote by e this energy arriving per s per cm^2, $e = \pi f$, where f is called the flux*. The energy arriving per cm^2 and per s per cm wavelength band at the wavelength λ is called πf_λ. Obviously, the total flux

$$\pi f = \int_0^\infty \pi f_\lambda \, d\lambda,$$

integrated over all wavelengths.

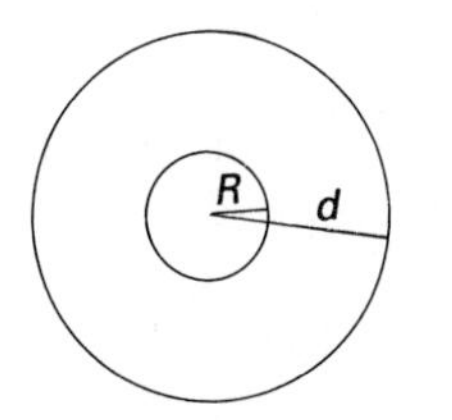

Fig. 6.1. The amount of energy leaving the star per s will pass through a sphere of radius d some time later. The time difference is given by the travel time between the star and the observer at distance d. The amount of energy passing through the sphere of radius d per s is the same as the amount of energy leaving the sphere of radius R per s, namely the star.

* Most observers write $e = f$ and include the factor π in the definition of f. We prefer the definition given above for theoretical reasons which will become apparent in Volume 2.

The energy πf which we receive does, of course, depend on the type of star which we observe and on its distance. At any given moment the whole star emits a certain amount of energy which is given by the amount emitted per cm^2 on the stellar surface, which we call πF, multiplied by the number of cm^2 on the stellar surface, which is given by

$$\text{total surface area} = 4\pi R^2, \tag{6.1}$$

where R is the radius of the star. The total amount of energy emitted by the star per s is then given by

$$\text{E(total)/s} = \pi F \cdot 4\pi R^2 = L = \text{luminosity}. \tag{6.2}$$

The energy emitted by the star per second is called the luminosity L of the star. After a time $t = (d - R)/c$ this same amount of energy will pass through the sphere at distance d. If there is no energy absorbed on the way, for instance by interstellar material, then the amount of energy passing through the sphere at distance d must be the same as was emitted by the star t seconds earlier. This leads to the equation

$$L = 4\pi R^2 \cdot \pi F = 4\pi d^2 \cdot \pi f \tag{6.3}$$

if the luminosity of the star does not change in time. Otherwise (6.3) describes the luminosity at the time $t = (d - R)/c$ before we observe the flux πf. (As usually $d \gg R$ we can replace $d - R$ by d). From (6.3) we then derive

$$\pi F = \pi f \cdot (d/R)^2. \tag{6.4}$$

We can determine the luminosity L by measuring the amount of energy arriving from the star above the Earth's atmosphere if we know the distance. We can determine the amount of energy leaving the stellar surface per cm^2 and per s if we measure πf and if we determine the ratio R/d, which is the angular radius of the star (see Fig. 6.2), if we measure the angular radius in radians. It is important to notice that it is often easier to determine πF the surface flux on the star than it is to determine the luminosity, because for

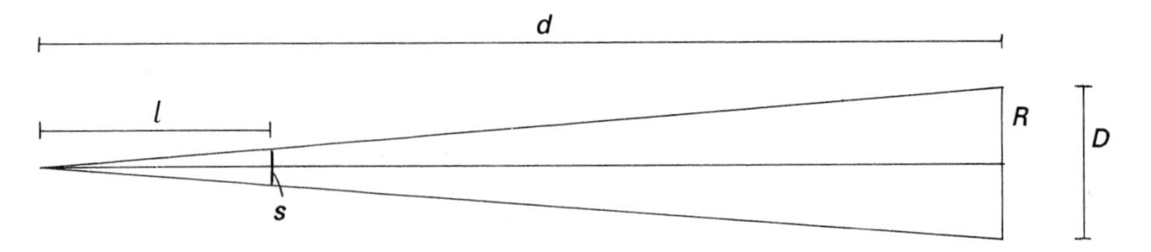

Fig. 6.2. The ratio R/d is the angular radius of the star as measured by the observer at distance d. A pencil piece of length s at a distance l from the observer covers the diameter of the sun. The ratio s/l can easily be measured, and equals $2R/d$.

stars not too distant it is easier to measure the angular diameter than it is to measure their distance.

6.2 The luminosity of the sun

The sun is close enough for us to see its disk and not just an unresolved dot of light as we see with our eyes for the other stars. For the sun everyone of us can therefore easily measure the angular diameter. An easy way to do so is to take a pencil in your hand, stretch out your arm and see what length s of the pencil will cover up the diameter of the sun (see Fig. 6.2). As can be seen from Fig. 6.2 the ratio of s to the length of your arm l gives you the angular diameter of the sun, which turns out to be about $D/d \approx 1/100$ (actually 0.0093). It is a very fortunate accident that the angular diameter of the moon is almost exactly the same, otherwise we would never be able to see total solar eclipses.

Knowing that the flux πf from the sun equals the measured value for the solar constant $S = 1.38 \times 10^6\,\mathrm{erg\,cm^{-2}\,s^{-1}} = \pi f(\mathrm{sun})$ we find with $R/d = D/(2d) = 1/2$ the solar angular diameter that

$$\pi F = S \cdot 200^2 \approx 6 \cdot 10^{10}\,\mathrm{erg\,cm^{-2}\,s^{-1}} \tag{6.5}$$

for the sun.

For the determination of the solar luminosity we have to know the distance to the sun, which, as we saw, is much more difficult to determine than the angular diameter, but which we know is $1\,\mathrm{au} = 1.49 \times 10^{13}\,\mathrm{cm}$. Inserting this into (6.3) we obtain for the sun

$$L(\mathrm{sun}) = 3.96 \times 10^{33}\,\mathrm{erg\ per\ s}. \tag{6.6}$$

This is a big number which really does not tell us a lot. It means more to us when we look at the energy generation of the sun. Knowing that the sun loses per s the energy L and seeing that it does not change with time, we must conclude that the sun must replenish the energy loss by some sort of energy generation. We can say that the energy generation per s equals $L = 3.96 \times 10^{33}$ erg. We understand this number better if we express L in kW, $1\,\mathrm{erg} = 10^{-10}\,\mathrm{kW}$, so $L(\mathrm{sun}) = 4 \times 10^{23}\,\mathrm{kW}$. From the orbits of the planets around the sun and from Kepler's third law we can determine the mass of the sun and find (see Section 9.5) that $M_\odot = M(\mathrm{sun}) = 1.98 \times 10^{33}\,\mathrm{g}$. (If we measure the luminosity of the sun in $\mathrm{erg\,s^{-1}}$ and the mass in g the ratio of $L/M = 2$. This can be remembered easily.) If we now calculate how much energy the sun produces per g of material we are surprised how little this is, namely 4 kW for each 2×10^{10} g of material, or 2 kW for each 10 000 tons of material. If we were to imitate the solar energy generation to cover our energy needs it would not do us much good, we need to do it much more efficiently. The sun has such an enormous mass that it can afford to be very inefficient and still be a very large energy source.

6.3 Luminosities of stars and bolometric magnitudes

If we measure the amount of energy which we receive from a star per cm^2 per s above the Earth's atmosphere we can determine the total amount of energy emitted by the star at all wavelengths if we also know the distance to the star. Actually we can only determine the luminosity for those wavelengths for which we receive energy above the Earth's atmosphere. Before we had rockets and satellites we could actually measure only the luminosity for those wavelengths which could still penetrate the Earth's atmosphere. But even now with satellites we do not get radiation from all wavelengths for most of the stars, because the interstellar material between us and the stars absorbs the light for wavelengths shorter than 912 Å for all the distant stars and absorbs a large fraction of that light even for the nearby stars. If stars emit a large fraction of their energy at such short wavelengths we cannot measure their luminosities directly, we have to use theoretical extrapolations to estimate the amount of energy emitted in those wavelength bands. We will see later (in Volume 2) how this can be done.

The luminosities or brightnesses of the stars integrated over all wavelengths are again measured in magnitudes, the so-called bolometric magnitudes. We can again distinguish apparent and absolute bolometric magnitudes, though generally only absolute bolometric magnitudes are in use. The difference between visual magnitudes m_V and bolometric magnitudes m_{bol} is called the bolometric correction BC. We then have

$$m_{bol} = m_V - BC \quad \text{or} \quad M_{bol} = M_V - BC. \tag{6.7}$$

Different authors disagree about the sign on the right-hand side of (6.7), which leads to different signs for the bolometric correction. This is of course only a matter of definition. Since the total radiation at all wavelengths is more than the radiation in a limited wavelength band, which means the bolometric brightness is larger than the visual, we expect that the *bolometric magnitudes* are *smaller* than the visual magnitudes. If we take the definition (6.7) with the – sign, then the BC are *generally positive*, though unfortunately not always for the following reasons. While all the apparent magnitudes for any wavelength band for Vega are 0 this is *not* the case for the apparent *bolometric* magnitude. For the bolometric magnitudes astronomers agreed on a different normalization. As we said the bolometric magnitudes measure the radiation for all wavelengths including the infrared and the ultraviolet region. For stars which have a large fraction of their radiation in those wavelength regions the bolometric corrections must be large. From the energy distribution of the black body radiation we know that hot objects emit a large fraction of their radiation at very short wavelengths while cool objects emit a large fraction in the infrared. We therefore expect the bolometric corrections to be large for very hot and very cool stars. There must be a minimum for intermediate temperatures. In Fig. 6.3 we have plotted the bolometric corrections for main sequence stars as a function of

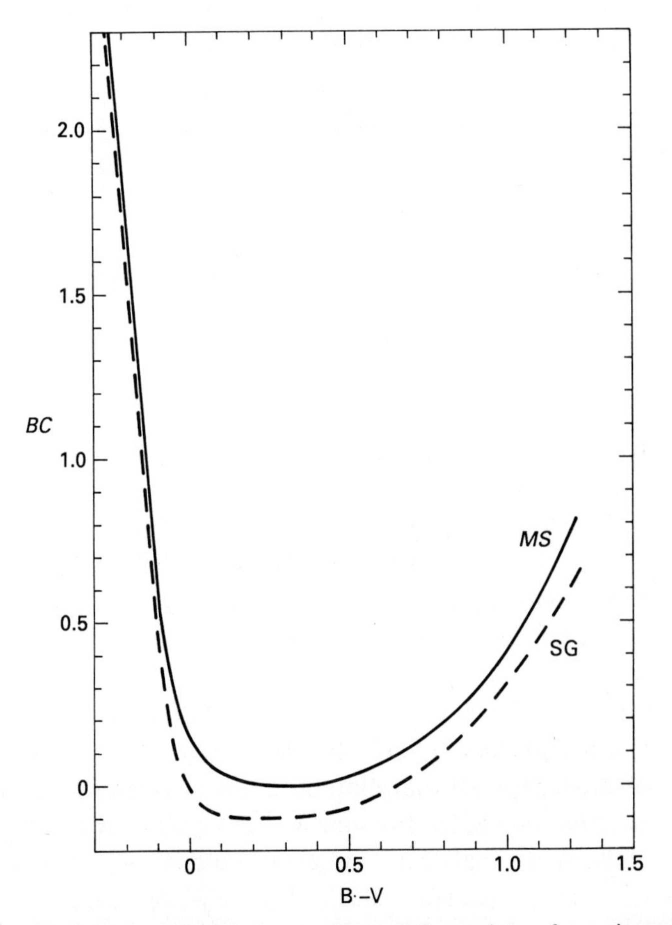

Fig. 6.3. The bolometric corrections, *BC*, are shown for main sequence stars with different B − V colors (solid line). These bolometric corrections have a minimum at B − V = 0.3. For main sequence stars the minimum bolometric correction was set equal to zero by definition. This now requires a change of sign of the *BC* for some supergiant stars as shown by the supergiant curve in this diagram (dashed curve).

Table 6.1. *Effective temperatures* T^{1}_{eff} *and bolometric corrections BC for main sequence stars and supergiants* $M_{\text{bol}} = M_{\text{V}} - BC$.

B − V	T_{eff} Main sequence	BC Main sequence	T_{eff} Super giants	BC Super giants
−0.25	24500	2.30	26000	2.20
−0.23	21000	2.15	23500	2.05
−0.20	17700	1.80	19100	1.72
−0.15	14000	1.20	14500	1.12
−1.10	11800	0.61	12700	0.53
−0.05	10500	0.33	11000	0.14
0.00	9480	0.15	9800	−0.01
+0.10	8530	0.04	8500	−0.09
+0.2	7910	0.00	7440	−0.10
+0.3	7450	0.00	6800	−0.10
+0.4	6800	0.00	6370	−0.09
+0.5	6310	0.03	6020	−0.07
+0.6	5910	0.07	5800	−0.03
+0.7	5540	0.12	5460	+0.03
+0.8	5330	0.19	5200	+0.10
+0.9	5090	0.28	4980	+0.19
+1.0	4840	0.40	4770	+0.30
+1.2	4350	0.75	4400	+0.59

1. Effective temperatures will be discussed in Chapter 8.

the B − V colors. They have a minimum for B − V = 0.3. Astronomers decided a long time ago to set the minimum bolometric correction to zero by definition. For main sequence stars with B − V = 0.3 we therefore have by definition $m_{\text{bol}} = m_{\text{V}}$ and $M_{\text{bol}} = M_{\text{V}}$.

Vega has B − V = 0 so for Vega m_{bol} is not equal to m_{V} which means it is not equal to zero. For stars like Vega the bolometric correction $BC = +0.15$.

The reason for this irrational definition of the BC was the desire to have bolometric corrections as small as possible without getting negative values. It now turns out, however, that we find negative bolometric corrections for the supergiants. So the original reason for the choice of the zero point for the bolometric corrections no longer holds. It seems to be much better to define the zero point for the m_{bol} consistently with the zero point for all other apparent magnitudes, but again astronomers do not like to make changes.

With this definition of the zero point the bolometric correction for the sun becomes $BC(\text{sun}) = 0.07$, and the absolute bolometric magnitude of the sun becomes $M_{bol}(\text{sun}) = 4.75$. It seems easiest to remember this number and derive all other absolute bolometric magnitudes from the relation

$$M_{bol}(\text{star}) - M_{bol}(\text{sun}) = -2.5 \cdot \log \frac{L(\text{star})}{L(\text{sun})}.$$

Bolometric corrections derived from measurements and theoretical extrapolations are given in Table 6.1.

7

Angular radii of stars

7.1 The problem

For the study of the structure of the outer layers of a star and also for the overall structure of the star it is very important to determine the surface flux of the radiation for the stars, which we call F. We saw in the previous section that the amount of energy leaving 1 cm^2 of the stellar surface per s, πF, can be obtained from the amount of energy reaching 1 cm^2 per s above the Earth's atmosphere, provided we know the angular radius of the star. Basically there are four methods in use to determine angular radii: the Michelson interferometer, the intensity interferometer, or Hanbury Brown interferometer, and lunar occultations. Speckle interferometry has also been used.

The basic difficulty with measuring stellar diameters is the fact that the angular diameters are so small that the blurring of the stellar image, which occurs when the light passes through the atmosphere, completely wipes out the size of the image. This blurring of the image is called the 'seeing' and is due to the turbulence elements in the air, which have slightly different temperatures and densities and therefore slightly different refractive indices. The light beams passing through different turbulence elements at different times therefore do not arrive exactly parallel at the telescopes but seem to come from slightly different directions and will be focused by the telescope at slightly different points. Since during the time of exposure of a photographic plate different turbulence elements pass by the telescope the images due to the different turbulence elements are all registered by the photographic plate. The sum of all these images then gives a rather large image of the star ranging from image sizes of about 1/2 arcsec for excellent seeing to 20 or 30 arcsec for very bad seeing. This means that even for the best seeing conditions stellar images of less than 1/2 arcsec cannot be resolved with ground-based telescopes which have to observe through the Earth's atmosphere. If we could observe for very short instances, when all the light beams from the star reaching the telescope pass through the same turbulence elements, we would be able to resolve much smaller images, but we would not receive enough light in such a short time. The interferometers circumvent this problem by comparing at any instance two beams of light which arrive at the telescopes simultaneously and have therefore in general passed through the same turbulence elements. The beams may change their directions slightly in time but they do so together, and that does no harm.

When the Space Telescope is in orbit we will be able to resolve images of 0.02 arcsec. Also the astrometric satellite Hipparchos will be able to resolve images of such small separation.

7.2 The Michelson interferometer

The Michelson interferometer was the first instrument in use to determine angular diameters of stars. Fig. 7.1 shows how the Michelson interferometer works. Two parallel light beams from the star enter the telescope lens at points A and B, where we have to provide two holes in a screen for the light to enter. At the focal point of the lens the beams form an image at the point P_1, where we see the image of the star. Since at points A and B the light enters through small holes there is light scattered at the edges of these holes. This scattered light is dispersed in all directions and appears on the focal plane all around point P_1. Scattered light comes from both holes, but the light coming from point A arrives at point P with a different phase from the light coming from point B because of the different path lengths. If at a point P the path length difference is $\lambda/2$ the phases differ by π and the electric and magnetic vectors cancel, there will be no light at the point P. On the other hand, if the path length difference is λ the phase difference is 2π and the electric and magnetic vectors add up, we see another, secondary light maximum at $P = P_2$. Another light maximum occurs at point $P = P_3$ (not plotted) for which the path length difference is 2λ. From Fig. 7.1 we see that the path length difference $\Delta s = \gamma D$, where D is the distance between the entrance holes A and B. For light maxima we require $\Delta s = \lambda$, for light minima $\Delta s = \lambda/2$. The distance l between points P_1 and P_2 is given by $l = \gamma d$ where d is the focal length.

For light maxima we therefore require

$$\gamma_{\max} = \lambda/D, \tag{7.1}$$

for light minima

$$\gamma_{\min} = \lambda/2D. \tag{7.2}$$

Let us now look at an image of a star 2 which is a small angular distance δ from star 1. The primary image of star 2 appears in the focal plane at a distance δ from the primary image of star 1. The image of star 2 also shows secondary maxima just the same as those from star 1. If the angular distance

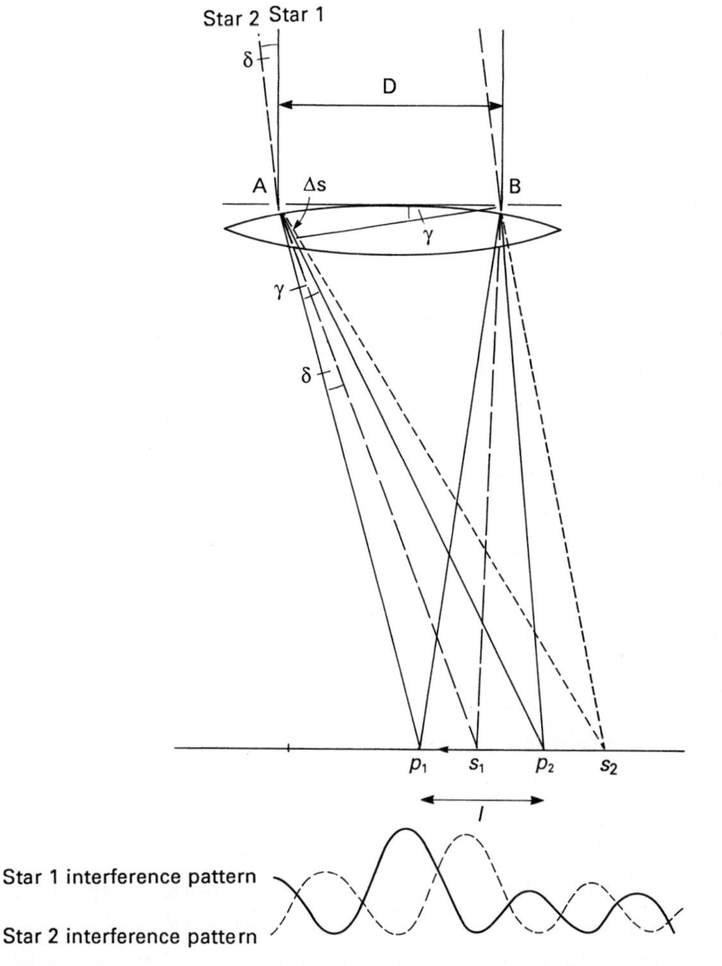

Fig. 7.1. A schematic plot of the light paths in the Michelson interferometer is shown. Light from star 1 enters through the entrance slits at points A and B. The primary image of star 1 is formed at point P_1 in the focal plane. The light scattered at the entrance slits forms an interference pattern around P_1. Secondary maxima occur around points P_2, P_3 etc., where the path length differences for light from A and B are multiples of the wavelength λ of the light. Light minima occur halfway between these points. Light from a star 2, which is at an angular distance δ from star 1 forms a similar interference pattern around its primary image S_1. If the light maxima from star 2 fit into the light minima of star 1 the interference pattern in the focal plane becomes invisible if both stars are equally bright. In this case (7.3) holds, from which δ can be determined.

δ of star 1 from star 2 happens to be

$$\delta = \gamma_{\min} = \lambda/2D \tag{7.3}$$

then the primary image of star 2 will appear at the place of minimum light for star 1 and the subsequent light maxima for star 2 all fall on the subsequent light minima for star 1. This means that the interference pattern of star 1 will be wiped out by the interference pattern of star 2, provided both stars are equally bright. If this happens we know that the angular distance between star 1 and star 2 is given by (7.3). We have then actually measured the angular distance δ of star 1 from star 2. Given a distance δ between two neighboring stars we can adjust the distance D between the two entrance holes in such a way that the interference patterns for the two stars becomes invisible, we then know that (7.3) must hold. We can measure the distance D for which this happens and then calculate δ from (7.3).

How can this help us to measure angular radii of stars? In our minds we can subdivide the star into two half stars (see Fig. 7.2), half 1 and half 2. The two half stars are now separated by approximately the angular radius of the star $\theta/2 = R/\text{distance}$. Knowing the brightness distribution on the stellar surface we can determine the distance of the 'centres of gravity' for the two halves more accurately. We can then make the same kind of observation as described above for the two stars. When the interference pattern, which is now observed for just one star at the focal plane, becomes invisible we know that $\delta = \gamma_{\min} = \lambda/2D$. We can again adjust D so that this happens. Instead of taking one very big lens with a screen in front with two openings, it is easier to take two separate lenses at a large distance apart, and then vary the positions of the two lenses until the interference pattern becomes invisible.

Due to the seeing effects the interference patterns of both stellar images will move back and forth on the focal plane, but they always move together, so the light maxima for star 2 always fall on the light minima for star 1, even

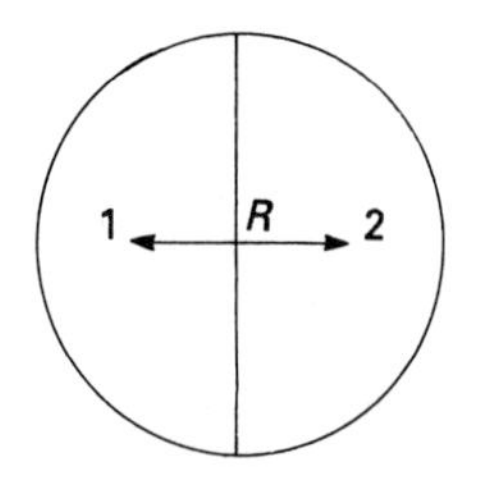

Fig. 7.2. In our minds we can divide the disk of a star into two halves, 1 and 2. The two halves are separated by approximately the angular radius of the star, which can thus be measured with the Michelson interferometer.

Table 7.1 *Angular diameters measured with the Michelson interferometer*

Star	Spectral[1] type	Distance D between lenses (meters)	Angular diameter Θ^2_{uD}
α Bootis	K0III	7.3	0.020
α Tauri	K5III	7.3	0.020
α Orionis	M0III	3.0	0.047 var.
β Pegasi	M5III	6.7	0.021
α Herculis	M8III	4.9	0.030
o Ceti	M7III	3.0	0.047 var.
α Scorpii	M0III	3.7	0.040

1. The spectral types will be discussed in Chapter 10.
2. This is the angular diameter which the star would have to have if its disk had uniform brightness. All stars are expected to be dimmer at the limb than in the center, as is observed for the sun (see Volume 2). The true angular diameter is therefore probably slightly larger.

though they are at different places on the screen. As long as the light beams from both stars go through the same turbulence elements the seeing does not disturb the observation. But this condition sets a limit to the distance D between the lenses, if D is too large then the light beams going through the two lenses pass through different turbulence elements and have additional phase differences which disturb the interference pattern. This upper limit on D sets a lower limit on the distance δ and on the angular radius which can be measured. Only angular radii of very large stars can be measured in this way. In Table 7.1 we give some values for angular radii of stars which have been measured in this way.

7.3 The Hanbury Brown interferometer

The Hanbury Brown interferometer works on a very different principle. It also uses two large mirrors but interferences are measured for each mirror alone and then the interferences for these mirrors are compared (see Fig. 7.3). If the light falling on these two mirrors is coherent, then there is a good correlation between the interferences measured for the two telescopes. How can there be interferences in each of the separate mirrors? Observations are made in a narrow wavelength band and the light waves of slightly different wavelengths show interferences, which are also called beat phenomena. The amplitude of a wave with a given frequency itself has a wave

pattern. In Fig. 7.4 we show the beat phenomenon on a wave pattern resulting from the superposition of two waves of slightly different frequencies. The frequency of the amplitude modulation is given by the difference in the frequencies of the two interfering waves:
Writing

$$2\cos\omega t = e^{i\omega t} + e^{-i\omega t}$$

we find quite generally,

$$\cos\omega_1 t + \cos\omega_2 t = 2\cos\frac{\omega_1 - \omega_2}{2}t\cdot\cos\frac{\omega_1 + \omega_2}{2}t, \tag{7.4}$$

which describes a wave with a frequency $(\omega_1 + \omega_2)/2$ with an amplitude which is modulated with the frequency $(\omega_1 - \omega_2)/2$.

Different wave amplitudes of a light wave mean different intensities. Due to this beat phenomenon light fluctuations are observed in each telescope over very short time scales. With the Hanbury Brown interferometer intensities therefore have to be measured at very short time intervals in order

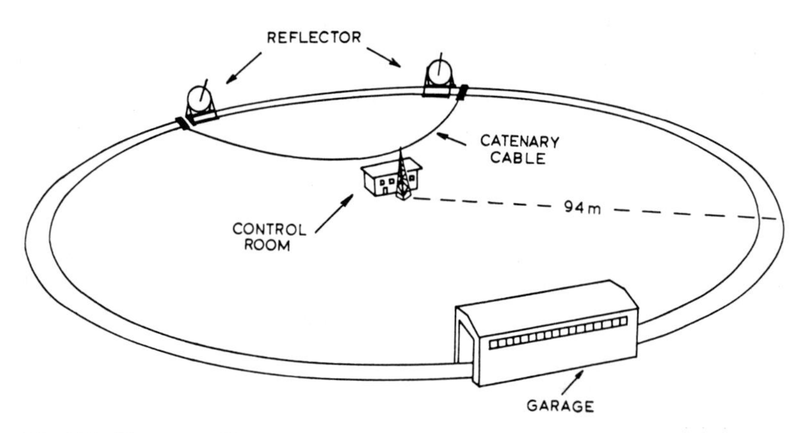

Fig. 7.3. The general layout of the Hanbury Brown interferometer is shown. (From Hanbury Brown, 1974.)

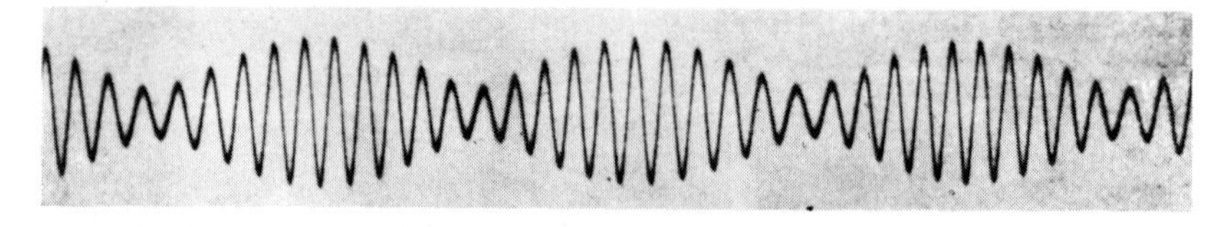

Fig. 7.4. The beat phenomenon resulting from the superposition of two waves of nearly equal frequencies is shown. The amplitude of the resulting wave is itself a wave.

to measure these intensity fluctuations. In order to get measurable energies over such short time scales large mirrors have to be used and also bright stars are needed for the measurements. As we shall see soon, the stars must not be too large in order to be measurable with this type of instrument, otherwise the mirrors would have to be too small and could not gather enough light.

In order to understand how these intensity fluctuations can be used to measure angular distances between nearby stars or to measure angular radii of stars we have to give a clear definition of what is meant by coherent light. In the strict definition light is coherent or two photons are coherent if they belong to the same quantum cell in phase space. The phase space element is given by

$$\mathrm{d}(\text{phase space}) = \mathrm{d}x\,\mathrm{d}y\,\mathrm{d}z\,\mathrm{d}p_x\,\mathrm{d}p_y\,\mathrm{d}p_z$$

x, y, z are the space coordinates and p_x, p_y and p_z are the momentum coordinates in the directions of x, y and z. For coherent light the two photons being coherent have to be in a phase space element d(phase space) which is only one quantum cell, which means it cannot be larger than h^3, where h is the Planck constant. The intensity fluctuations measured in the two mirrors are correlated as long as the light falling on the two mirrors is coherent. From the schematic representation of the Hanbury Brown interferometer in Fig. 7.5 we see the following relations. We put the coordinate system in such a way that the x coordinate is in the direction from

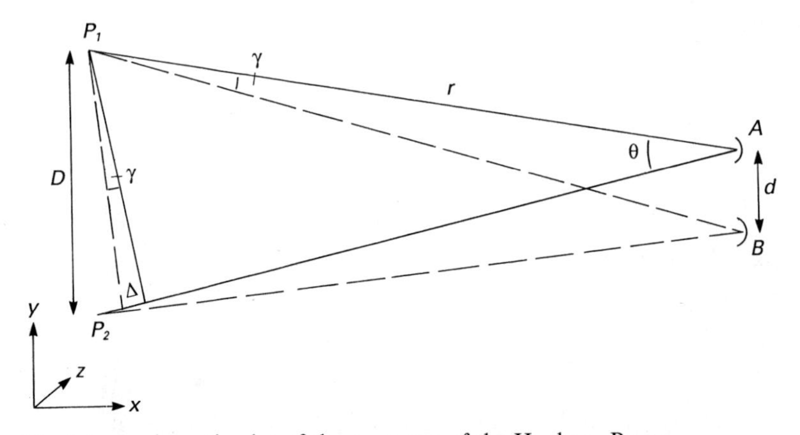

Fig. 7.5. A schematic plot of the geometry of the Hanbury Brown interferometer is shown: d is the distance between the two mirrors, D the distance between two points near the limb of the star. θ is the angular diameter of the star, and r the distance to the star. γ is the angular distance of the mirrors as seen from the star. The orientation of our coordinate system is also shown.

the star to the interferometer. The y and z directions are in the plane perpendicular to this direction. The photons reaching the mirrors at any given time come from any points P_1 and P_2 on the star, which means they can, in the y and z directions, have distances of the order of the stellar diameter D. This means $\Delta y = \Delta z = D$. If they arrive at the mirrors at the same time they must have had a distance Δx equal to the path length Δ, which means $\Delta x = \Delta$. From Fig. 7.5, we see that $\Delta = D\gamma$ and $\gamma = d/r$, where d is the distance between the two mirrors and r the distance of the star.

We now look at the differences in the momenta p_x, p_y, p_z of the photons reaching the mirrors. The main direction of propagation is the x direction, therefore $p_x = h\nu/c$, where ν is the frequency of the light observed. If a wavelength band of width $\Delta\nu$ is used, then the spread in p_x is $\Delta p_x = h\Delta\nu/c$. The photons reaching mirrors A and B must have a difference in $\Delta p_y = p_x\gamma$ where $\gamma = d/r$. The same holds for Δp_z.

We thus find for the volume element in phase space populated by the photons which reach the two mirrors at any given time

$$\Delta x\,\Delta y\,\Delta z\,\Delta p_x\,\Delta p_y\,\Delta p_z = \Delta D^2 \frac{h\Delta\nu}{c}\left(\frac{h\nu}{c}\gamma\right)^2 = \Delta V\,\Delta P \qquad (7.5)$$

If we observe the photons in the two mirrors to be coherent they must all belong to the same quantum cell, which means we must conclude that

$$\Delta V\,\Delta P < h^3 \qquad (7.6)$$

According to (7.5) this means

$$\Delta D^2 \frac{h^3\nu^3}{c^3}\frac{\Delta\nu}{\nu}\gamma^2 \leqslant h^3 \qquad (7.6a)$$

With $\Delta = D\gamma = Dd/r$ we find that for coherence we must have with $\lambda = c/\nu$ and after dividing by h^3

$$D^3(d/r)^3\lambda^{-3}\Delta\nu/\nu \leqslant 1 \qquad (7.6b)$$

or, taking the third root we find

$$d \leqslant \frac{r}{D}\lambda\sqrt[3]{\frac{\nu}{\Delta\nu}} = \Theta^{-1}\lambda\sqrt[3]{\frac{\nu}{\Delta\nu}} \qquad (7.7)$$

where we have made use of the relation that the angular diameter $\Theta = D/r$.

For larger distances d than given by (7.7) the photons reaching the two mirrors are not coherent any more because the p_y and p_z components of their momenta differ too much. Starting at small distances d between the mirrors and slowly increasing d we expect first to measure a good correlation between the intensity fluctuations in the two mirrors. Once we reach the distance d for which the equal sign holds in (7.7) we expect the correlation to decrease and finally go to zero for large distances of the mirrors: If $\Delta = \Delta x$ becomes too large the photons reaching the two mirrors are not coherent any more. For which Δ does this happen? From (7.7) we see that for the equal sign we find

$$\Delta = D\frac{d}{r} = \lambda \sqrt[3]{\frac{\nu}{\Delta\nu}} \tag{7.8}$$

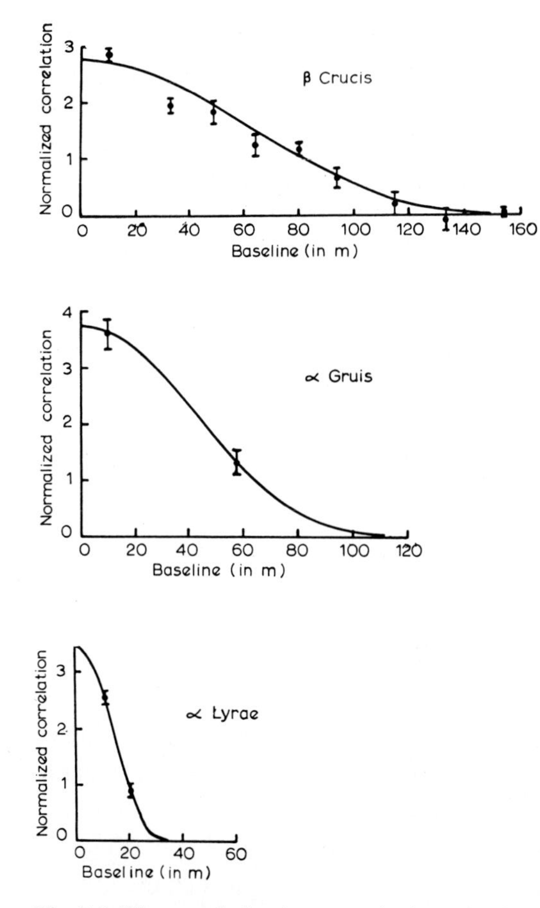

Fig. 7.6. The correlation between the intensity fluctuations of the two mirrors of the Hanbury Brown interferometer is shown as a function of distance between the mirrors. For large angular diameters, (see α Lyrae) the correlation breaks down for very small distances. For small angular diameters of the stars (for instance, β Crucis) the correlation keeps up for much larger distances. (From Hanbury Brown 1974.)

with $\Delta\nu/\nu = 10^{-2}$ and $\sqrt[3]{(\nu/\Delta\nu)} = 4.6$ we find that the coherence stops for $\Delta \approx 5\lambda$.

From (7.7) we see that large values of Θ can only be measured with small distances d of the mirrors, but the smallest distance of the mirrors is limited by the size of the mirrors. For the measurement of large Θ we would need very small mirrors which would, however, gather too little light. So very large angular diameters can not be measured with the Hanbury Brown interferometer, while small angular diameters can be measured well as long as the star is still bright enough to measure very short time (10^{-9} s) fluctuations. For small λ small angular diameters can be measured especially well. The Hanbury Brown interferometer works best for bright stars with small angular diameters, which means for hot luminous main sequence stars. It is therefore complementary to the Michelson interferometer with which only large angular diameters can be measured, which are mainly observed for cool supergiants.

In Fig. 7.6 we show for some stars the measured correlations between the fluctuations in the two mirrors as a function of distance between the mirrors. For large distances the correlation drops. The smaller the angular diameter, the smaller the distance for which the correlation decreases.

In Table 7.2 we give the angular diameters of stars as measured by Hanbury Brown.

7.4 Angular diameters from lunar occultations

For stars near the ecliptic, which means near the plane of the Earth's orbit, angular diameters of stars can also be measured by means of lunar occultations when the dark side of the moon covers up the star.

Since the moon orbits the Earth in 28 days it apparently goes around the sky once a month, it can eclipse stars which happen to be behind its orbit as seen by us. Because the moon's orbit is inclined with respect to the Earth's orbital plane by about 5° and because the Earth orbits the sun, different stars can be occulted at different times. Even though only stars in a small angular range about the ecliptic can be studied in this way a fairly large number of stellar angular diameters have been determined by lunar occultations. Of course, we always have to wait for a time when the stellar eclipse happens naturally and hope that at that particular time the weather is also cooperative.

In principle, we just have to measure the time between the start of the decline of the stellar light until the light is completely gone. The moon moves 360° in one month, or roughly in 3×10^6 s. This means that in 1 s of time it

Table 7.2 *Angular diameters, θ, measured with the Hanbury Brown interferometer*

Star[1] number	Star name	Type[2]	Angular diameter[3] in 10^{-3}s of arc $\Theta_{LD} \pm \sigma$	Temperature[4] $[T_{eff} \pm \sigma]$/K
472	α Eri	B3(Vp)	1.92 ± 0.07	$13\,700 \pm 600$
1713	β Ori	B8(Ia)	2.55 ± 0.05	$11\,500 \pm 700$
1790	γ Ori	B2 (III)	0.72 ± 0.04	$20\,800 \pm 1300$
1903	ε Ori	B0(Ia)	0.69 ± 0.04	$24\,500 \pm 2000$
1948	ζ Ori	O9.5(Ib)	0.48 ± 0.04	$26\,100 \pm 2200$
2004	κ Ori	B0.5 (Ia)	0.45 ± 0.03	$30\,400 \pm 2000$
2294	β CMa	B1 (II-III)	0.52 ± 0.03	$25\,300 \pm 1500$
2326	α Car	F0 (Ib-II)	6.6 ± 0.8	7500 ± 250
2421	γ Gem	A0(IV)	1.39 ± 0.09	9600 ± 500
2491	α CMa	A1(V)	5.89 ± 0.16	$10\,250 \pm 150$
2618	ε CMa	B2 (II)	0.80 ± 0.05	$20\,800 \pm 1300$
2693	δ CMa	F8 (Ia)	3.60 ± 0.50	...
2827	η CMa	B5 (Ia)	0.75 ± 0.06	$14\,200 \pm 1300$
2943	α CMi	F5 (IV-V)	5.50 ± 0.17	6500 ± 200
3165	ζ Pup	O5(f)	0.42 ± 0.03	$30\,700 \pm 2500$
3207	γ^2 Vel	WC8 + O9	0.44 ± 0.05	$29\,000 \pm 3000$
3685	β Car	A1(IV)	1.59 ± 0.07	9500 ± 350
3982	α Leo	B7(V)	1.37 ± 0.06	$12\,700 \pm 800$
4534	β Leo	A3(V)	1.33 ± 0.10	9050 ± 450
4662	γ Crv	B8(III)	0.75 ± 0.06	$13\,100 \pm 1200$
4853	β Cru	B0.5(III)	0.722 ± 0.023	$27\,900 \pm 1200$
5056	α Vir	B1(IV)	0.87 ± 0.04	$22\,400 \pm 1000$
5132	ε Cen	B1(III)	0.48 ± 0.03	$26\,000 \pm 1800$
5953	δ Sco	B0.5(IV)	0.46 ± 0.04	...
6175	ζ Oph	O9.5(V)	0.51 ± 0.05	...
6556	α Oph	A5(III)	1.63 ± 0.13	8150 ± 400
6879	ε Sgr	A0(V)	1.44 ± 0.06	9650 ± 400
7001	α Lyr	A0(V)	3.24 ± 0.07	9250 ± 350
7557	α Aql	A7 (IV, V)	2.98 ± 0.14	8250 ± 250
7790	α Pav	B2.5(V)	0.80 ± 0.05	$17\,100 \pm 1400$
8425	α Gru	B7(IV)	1.02 ± 0.07	$14\,800 \pm 1200$
8728	α PsA	A3(V)	2.10 ± 0.14	9200 ± 500

1. Bright star catalog number (Hoffleit and Jaschek, 1982).
2. Spectral type and luminosity class (in brackets), to be discussed in Chapter 10.
3. True angular diameter allowing for the effects of limb-darkening.
4. Effective temperatures will be discussed in Chapter 8.

moves about 0.5 arcsec. If the star has an angular diameter of 10^{-3} arcsec, then we have to have a time resolution of the order of 10^{-3} s or better. During such short time intervals we must be able to measure the radiation flux of the star. This again means that the stars have to be bright and we have to use large telescopes in order to gather enough light in 10^{-3} s.

In reality the stellar light does not disappear monotonicly, but the light rays from the star are diffracted at the moon's limb (Fresnel diffraction). The superposition or the 'interference' of the different diffracted beams, originating on different parts of the star, results in an interference pattern (see Fig. 7.7, where two such calculated interference patterns are reproduced). The interference pattern depends on the angular diameter of the star. The measured pattern has to be compared with the ones calculated for different angular diameters. The one which fits the observation best tells us the

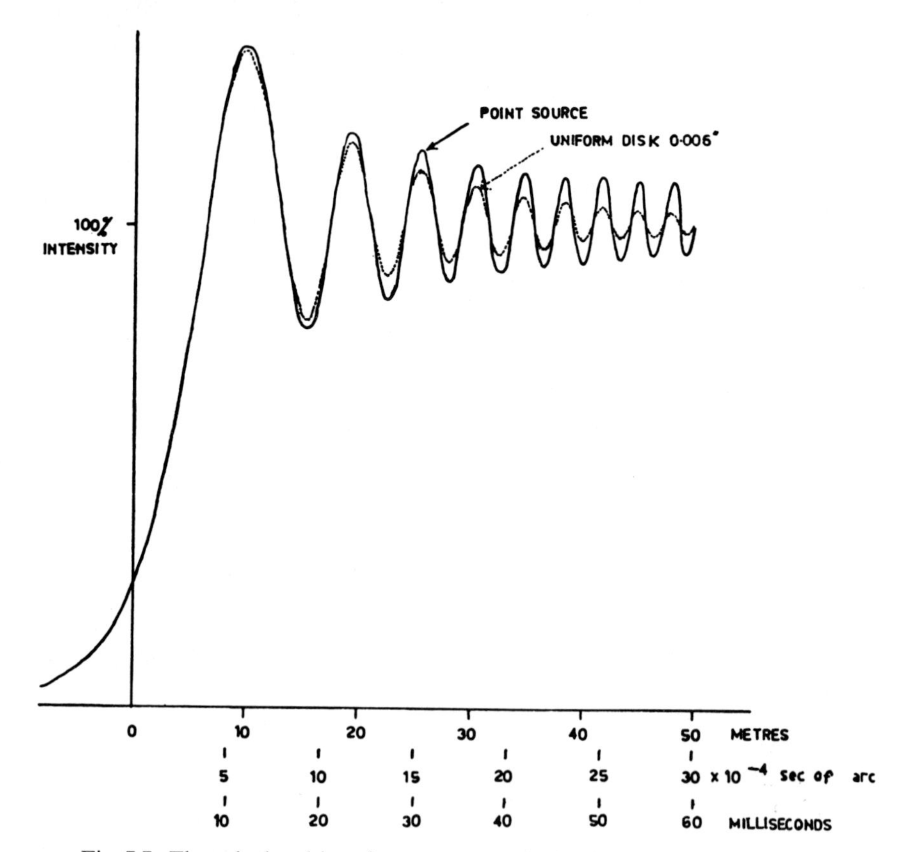

Fig. 7.7. The calculated interference pattern during lunar occultations is shown for a point source. The three scales of the abscissae represent the size of the pattern on the Earth in meters, the angle subtended by this pattern at the moon in seconds of arc, and the variation of light with time at a fixed point in milliseconds. (From Hanbury Brown 1968.)

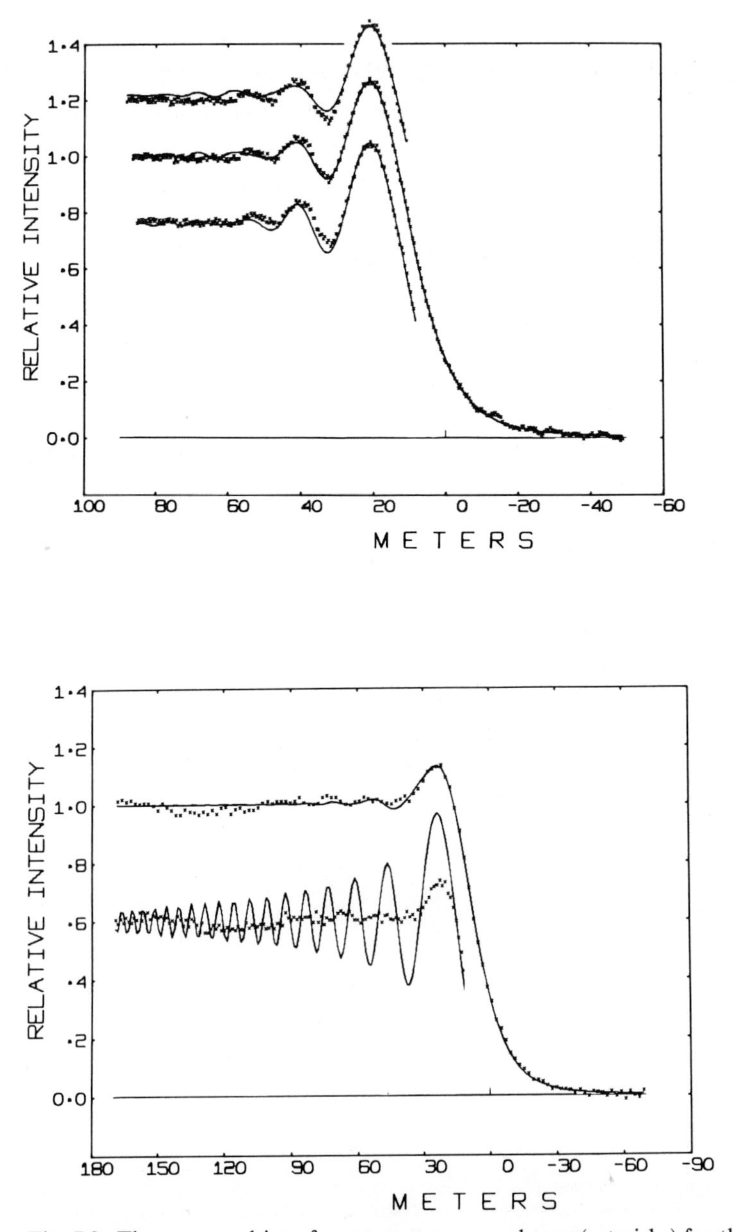

Fig. 7.8. The measured interference patterns are shown (asterisks) for the lunar occultation of Y Tau (top) and for U Ori (bottom). Also shown are the theoretical patterns for an angular diameter of $\Theta = 7.98$ milli-arcsec (center curve for Y Tau), which is considered to give the best fit for Y Tau, and the theoretical curves for $\Theta = 8.98$ milli-arcsec (top graph) and for $\Theta = 6.98$ milli-arcsec (bottom curve for Y Tau). In the plot for U Orionis the theoretical curve for $\Theta = 15.45$ milli-arcsec is shown, which is considered to be the best fit curve (upper curve), and also the one for a point source (lower curve). (From Ridgway, Wells, & Joyce 1977.)

angular diameter of the star. In Fig. 7.8 we reproduce some of the observed patterns and the best fit calculated ones.

In Table 7.3 we give some of the stellar angular diameters determined from lunar occultation measurements.

Table 7.3 *Some angular diameters measured from lunar occultations according to Ridgeway, Wells and Joyce (1977)*

HR No.[2]	Star	Sp. type[3]	B − V	mv	Θ_{LD}[1] 10^{-3} arcsec	π arcsec	$R/R_\odot$	T_{eff}
2286	μ Gem	M3III	1.64	2.88	13.65	0.020	73.2	3650
867	RZ Ari	M6III	1.47	5.91	10.18	0.014	78.0	3160
1977	Y Tau	C5II	3.03	6.95	8.58	?	?	?
5301		M2III	1.72	4.91	4.41	0.012	39.4	4040[4]
4902	ψ Vir	M3III	1.60	4.79	5.86	0.021	29.9	3530
7150	ξ^2 Sgr	K1III	1.18	3.51	3.80	0.011	37.1	4210
9004	TX Psc	C5II	2.60	5.04	9.31	−0.004	?	?
3980	31 Leo	K4III	1.45	4.37	3.55	?	?	4000
7900	υ Cap	M2III	1.66	5.10	4.72	0.019	26.6	3440

1. Considering the limb darkening of the disk.
2. Number in the Bright Star catalogue, Hoffleit and Jaschek (1982).
3. From the Bright Star catalogue
4. Assuming BC = 2.0.

8

Effective temperatures of stars

8.1 General discussion

As we have seen in Chapter 6 (6.4), we can determine the surface flux πF per cm^2 and per s if we measure the flux πf arriving above the Earth per cm^2 and per s and if we know the angular diameter or angular radius R/d for the star. With $\theta = R/d$ we have

$$\pi F = \pi f \cdot 1/\theta^2. \tag{8.1}$$

The surface flux tells us how much radiation is emitted from 1 cm^2 per s by the star. In order to get another estimate for the temperature of a star we can compare this surface flux of the star with the flux emitted by a black body of temperature T which is, according to (4.5),

$$\pi F(\text{black body}) = \sigma T^4. \tag{8.2}$$

If the star did radiate like a black body we could use this equation to determine its temperature. We know, however, that the stars are not black bodies because they are not well insulated from the surroundings, they actually lose a lot of energy all the time to the surroundings, fortunately for us, because otherwise we would not be able to see them and we would have no knowledge of their existence.

So we know that the stars are not black bodies but nevertheless we can compare their energy loss per cm^2 with that of a black body and say that if they did radiate like black bodies then their temperatures would have to be given by

$$\sigma T^4 = \pi F = \sigma T_{\text{eff}}^4. \tag{8.3}$$

The temperature defined in this way is called the *effective temperature* of the star. It is another way of describing the surface flux of a star. All it means is that the energy loss of the star per cm^2 and per s is the same as that of a black body with $T = T_{\text{eff}}$.

If we measure the black body radiation coming out of the hole in the box, that means the black body is permitted to lose energy by radiation it is then, of course, in the strict sense, not a black body any more because it is no longer insulated from the surroundings. We have to make sure that the hole in the box through which the radiation escapes is very small, such that the energy loss remains very small in comparison with the energy content of the black body. This consideration also gives us hope that, may be, the stars are not so different from a black body after all because their energy loss, large as it may be, may still be small compared to their energy content.

Equation (8.3) is the definition of the effective temperature of a star. In Volume 2 we will see that this T_{eff} is indeed the temperature in the surface layer of the star from which the radiation which we receive is mainly emitted. We can consider the effective temperature as an average temperature of the surface layers of the star.

In Chapter 7 we have discussed how angular radii can be measured for several groups of stars. For these stars we can directly determine T_{eff} *if* we know πf. The determination of πf requires, however, a knowledge of the stellar radiation at *all* wavelengths including those which cannot penetrate the Earth's atmosphere.

8.2 The solar surface flux and effective temperature

The sun fortunately does not emit much radiation in the infrared and ultraviolet wavelengths which cannot penetrate the Earth's atmosphere. We can now even measure the total radiation of the sun from satellites above the Earth's atmosphere and find

$$\pi f(\text{sun}) = S = 1.38 \times 10^6 \text{ erg cm}^{-2} \text{ s}^{-1}. \quad (8.4)$$

With the known angular diameter of the sun (see Section 6.2), namely $\Theta = 0.0093$ radians, we calculate

$$\pi F(\text{sun}) = 6.3 \times 10^{10} \text{ erg cm}^{-2} \text{ s}^{-1}. \quad (8.5)$$

With $\pi F(\text{sun}) = \sigma\, T_{eff}^4(\text{sun})$ we calculate

$$T_{eff}(\text{sun}) = 5780 \text{ K} \quad (8.6)$$

It is very helpful to remember this value as a reference.

We remember that we estimated the solar temperature to be about 6000 K from the wavelength λ_{max} for which we observe the maximum of the solar energy distribution, using Wien's displacement law, (4.7). The close agreement of this value with T_{eff} for the sun gives us hope that these values do actually tell us something about the temperature of the outer layers of the sun, and that the solar radiation is actually not so different from that of a black body. In Volume 2 we will specify what we mean by 'not so different.'

8.3 Effective temperatures of stars

For stars more blue than the sun, which we expect to be hotter than the sun, we expect more radiation at the short wavelengths, i.e. in the ultraviolet. With the astronomical satellites we can now measure radiation for all wavelengths. In principle we could therefore determine the f_λ for all wavelengths. There is however one basic difficulty for which there is no straightforward solution: the interstellar material which is between us and the stars has a very large absorption coefficient for light with wavelengths shorter than 912 Å (see Chapter 17). We do not get any radiation in these wavelengths for stars further away than say about 50 pc. Also for wavelengths longer than 912 Å there is already rather strong absorption for stars with distances larger than about 200 pc. For hot stars we can therefore only determine πf accurately for stars which are closer than about 20 pc. There are no very blue stars that close. We therefore do not have accurate knowledge of T_{eff} for very hot stars, though we have some good estimates

Table 8.1 *Adopted values of T_{eff} for stars with different $B-V$ colors and different luminosity classes (The uncertainty is at least 5%)*

B – V	V[1]	III[2]	I[3]	B – V	V	III	I
–0.32	37000	37000		0.12	8380	8370	8000
–0.31	33000	33000	35000	0.16	8150	8130	7700
–0.30	30500	30500	34700	0.20	7910	7890	7440
–0.29	28700	28700	34400	0.30	7300	7270	6800
–0.28	27200	27200	33800	0.40	6750	6710	6370
–0.27	25700	25700	32500	0.50	6310	6270	6020
–0.26	24500	24500	26000	0.60	5910	5860	5740
–0.24	22000	22000	23500	0.70	5540	5480	5460
–0.22	19800	19800	21200	0.80	5330	5260	5220
–0.20	17700	17700	19100	0.90	5090	5020	4980
–0.16	14500	14500	15700	1.00	4840	4780	4770
–0.12	12500	12500	13500	1.10	4600	4580	4570
–0.08	11100	11100	11900	1.20	4350	4300	4410
–0.04	10200	10200	10700	1.30	4100	4060	4260
0.00	9480	9480	9600	1.40	3850	3900	4140
+0.04	9080	9080	9000	1.50	(3500)	3780	4020
+0.08	8700	8700	8500				

1. V stands for main sequence stars
2. III stands for giants
3. I stands for supergiants

from theoretical extrapolations from the observed part of the energy distribution. The most primitive method of extrapolation would be to determine an approximate T_{eff} and assume that the star would indeed radiate like a black body with $T = T_{\text{eff}}$. The Planck function then tells us how much radiation there should be in the short wavelengths, for which we cannot observe the flux. We have to add this amount to the observed energy and determine a better value for T_{eff}. This does not give very good corrections, because the hot stars do not radiate like black bodies in the short wavelength region. In Volume 2 we will learn how to do it better.

In Table 7.2 we have given the T_{eff} for those stars whose angular diameters have been determined by the Hanbury Brown interferometer and in Table 7.3 we collected the data measured with lunar occultations. In Table 8.1 we give the values of T_{eff}, which we consider to be the best ones, as a function of the B – V colors for main sequence stars (V), for giants (III) and for supergiants (I).

These values refer to stars in our neighborhood, which we call population I stars. We will see later (Chapter 12) that there are other stars which do not fit into this population. For those we find another relation between colors and effective temperatures.

9

Masses and radii of stars

9.1 General discussion of binaries

For the stars for which angular diameters have been measured we can, of course, determine the radii by multiplying the angular radius by the distance to the star, provided we know the distance. For stars further away than about 20 pc we cannot measure trigonometric parallaxes accurately, we can therefore determine distances only in indirect ways. Star stream parallaxes (see for instance Becker, 1950) have been used to determine the distance to the Hyades cluster, but it turned out that the photometric parallaxes were more accurate. Fortunately, we can also determine radii for stars in special binary systems, namely in eclipsing binaries. For these binaries we can also determine the masses of the stars. We shall therefore devote this section to the discussion of binaries in general, and the next sections to the discussion of the special types of binaries for which stellar radii can be determined, namely eclipsing binaries, and for those binaries for which stellar masses can also be determined, again the eclipsing binaries and also the visual binaries.

We can determine stellar masses for binaries using Kepler's third law. Let us briefly discuss the mechanics of a binary system.

From the solar system we are accustomed to the case that one body, the sun, has a much larger mass than the other bodies, the planets. For binaries with two stars we have to take into account that both bodies have elliptical (or perhaps circular) orbits around their center of gravity S (see Fig. 9.1). From the definition of the center of gravity we know that

$$M_1 r_1 = M_2 r_2 \tag{9.1}$$

where M_1 and M_2 are the masses and r_1 and r_2 are the distances of the stars from the center of gravity. In a binary system there must be equilibrium between gravitational and centrifugal forces which means

$$\frac{GM_1 M_2}{(r_1 + r_2)^2} = M_1 \omega_1^2 r_1 = M_2 \omega_2^2 r_2 \tag{9.2}$$

and where ω_1 and ω_2 are the angular velocities, r_1 and r_2 are the orbital radii of the two stars, and G is the gravitational constant. For the two stars to remain in phase, which means for the center of gravity to remain at constant velocity, we must have $\omega_1 = \omega_2 = \omega$.

For (9.2) to be true, we must then also have $M_1 r_1 = M_2 r_2$ which means that both stars must orbit the center of gravity.

For *circular* orbits ω can be replaced by

$$\omega = 2\pi/P \tag{9.3}$$

where P is the orbital period. We then derive

$$\frac{GM_1 M_2}{(r_1 + r_2)^2} = M_1 r_1 \frac{4\pi^2}{P^2}, \tag{9.4}$$

division by $M_1 r_1 4\pi^2$ yields

$$\frac{G}{4\pi^2} \cdot \frac{M_2}{r_1} \frac{1}{(r_1 + r_2)^2} = \frac{1}{P^2}. \tag{9.5}$$

Using (9.1) we find $M_1 = (M_2 \cdot r_2)/r_1$ and

$$(M_1 + M_2) = M_2\left(\frac{r_2}{r_1} + 1\right) \quad \text{or} \quad M_1 + M_2 = \frac{M_2}{r_1}(r_2 + r_1). \tag{9.6}$$

This yields

$$\frac{M_2}{r_1} = \frac{M_1 + M_2}{r_2 + r_1}. \tag{9.7}$$

Inserting (9.7) into (9.5) yields the general form of Kepler's third law

$$M_1 + M_2 = \frac{(r_1 + r_2)^3}{P^2} \cdot \frac{4\pi^2}{G}. \tag{9.8}$$

If distances are measured in astronomical units and times in years and masses in units of solar masses, the factor $4\pi^2/G$ becomes unity.

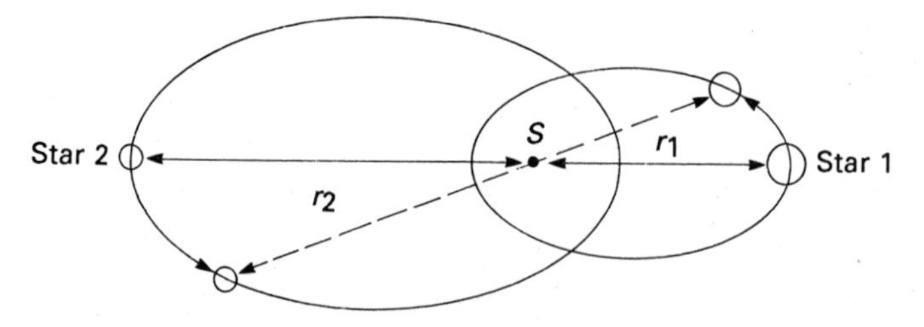

Fig. 9.1. In a binary system the two stars, Star 1 and Star 2 orbit their center of gravity, S.

While we have derived (9.8) only for circular orbits (otherwise r_1, r_2 and ω would be functions of time) it can be shown that this law also holds for elliptical orbits, if r_1 and r_2 are replaced by the semi-major axes a_1 and a_2 of the orbits around the center of gravity.

In order to determine the masses, we then have to determine a_1 and a_2, or at least the sum. If we look at visual binaries, i.e., binaries for which we can observe both stars, we will generally only be able to observe the orbit of one star relative to the other. For circular orbits of the two stars, the relative orbits of each star around the other will also be a circle with radius $r_1 + r_2$. Making a plot of the relative positions of one star around the other for elliptical orbits one can see that each star will also go around the other in an ellipse and the second star will be in the focal point of the true elliptical orbit.* The semi-major axis of the orbit of one star around the other is $a_1 + a_2$. This can be seen from Fig. 9.2 as follows.

The major axis of the relative orbit is

$$b_1 + b_2 + d_2 + d_1 = b_1 + d_1 + b_2 + d_2 = (a_1 + a_2)2 = 2a \tag{9.9}$$

which is the sum of the major axes of the two orbits around the center of gravity. We have written a_1 and a_2 for the semi-major axes of the stellar orbits around the center of gravity, and a for the semi-major axis of the relative orbit of one star around the other.

What we actually observe for a visual binary system is unfortunately not

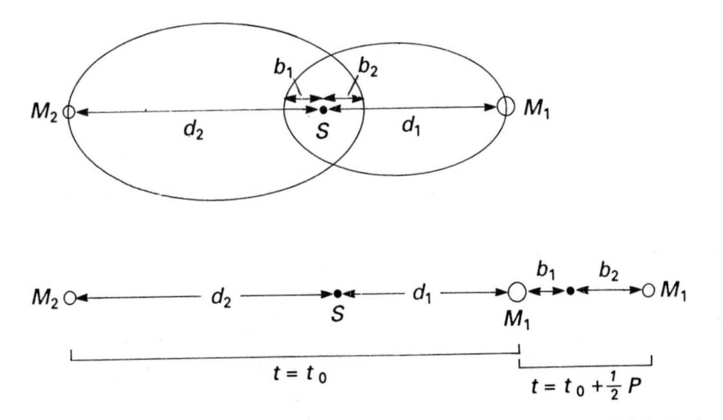

Fig. 9.2. The major axis of the relative orbit of M_1 around M_2 is the sum of the major axes of the orbit of each star around the center of gravity, S. t stands for time and P for the period, t_0 is the time of maximum distance apart of the two stars.

* What we see is, of course, generally not the true orbit but is the projection of the true orbit against the background sky. We will outline below how the true orbit can be determined from the projected orbit and the measured velocities.

the major axis itself but the angular major axis, namely $2a$/distance. In order to convert this into the actual major axis we have to determine the distance or find another way of conversion. As we know good parallaxes can only be determined for nearby stars. Fortunately the Doppler effect is independent of distance and provides a geometrical scale.

9.2 The Doppler effect

The Doppler effect is very important for many branches of astronomical research because it permits us to measure the component of the velocity along the line of sight, the so-called radial velocity (see Fig. 9.3). For the stars, we cannot of course measure their velocity by measuring the distance which they travel and then dividing by the time it takes them to do so, as we do for objects here on Earth. Remember that we can only observe the light from the star, from which we have to derive everything. Fortunately the frequency of the light that we see changes when the light source moves towards us or away from us. This shift in frequency or wavelength is called the Doppler effect. Fortunately the stellar spectra have very narrow wavelength bands in which they have very little light intensity, the so-called absorption lines, which mark distinct wavelengths whose Doppler shift can be measured. If the light source moves towards the observer the wavelength becomes shorter by $\Delta\lambda$ where

$$\frac{\Delta\lambda}{\lambda_0} = \frac{v_r}{c}, \tag{9.10}$$

here λ_0 is the so-called rest wavelength of the line (i.e., the wavelength which the line has, if the light source and the observer have the same velocity). v_r is the component of the velocity in the direction of the observer, the line of sight or the radial velocity (see Fig. 9.3).

If the light source moves away from us, the observers, the wavelength becomes longer by the same relative amount. It is this wavelength shift which permits us to measure the radial velocities of the stars in cm/s.

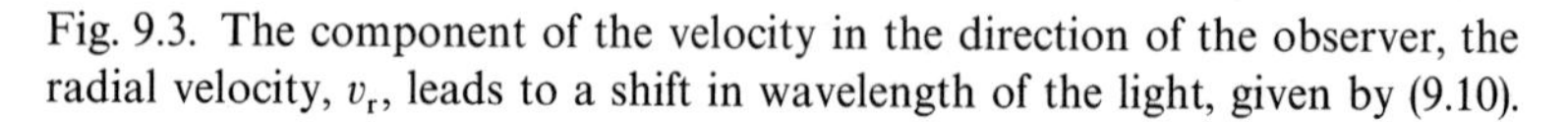

Fig. 9.3. The component of the velocity in the direction of the observer, the radial velocity, v_r, leads to a shift in wavelength of the light, given by (9.10).

9.3 Radial velocities and masses of binaries

In order to measure the radial velocities of the stars we have to observe the energy distributions of the stars in very narrow wavelength bands, i.e., we have to take so-called high ressolution spectra, such that we can see and measure the exact wavelengths of the spectral lines. If our line of sight happens to be in the orbital plane of the binary system, consisting of stars *A* and *B* orbiting in circular orbits around their center of gravity *S* (see Fig. 9.4), then at the time of largest angular distance of the two stars, star *B* will move towards us while star *A* moves away from us. If stars *A* and *B* have the same mass, as assumed in Fig. 9.4, then the velocities will be of the same magnitude but in opposite directions. At this time, the stars have the largest radial velocity component. While the speed of the two stars in their circular orbits remains the same, the component in the line of sight decreases when they move on and is zero when they reach points P_1 and P_2, where their motions are perpendicular to the line of sight. At this time, the spectral lines, also called Fraunhofer lines, appear at the wavelength λ_0 corresponding to the velocity of the center of gravity. The radial velocity components then change directions for both stars and the Doppler shift changes sign. The lines of both stars are now shifted in the opposite direction. Altogether the lines of each star show a periodic shift whose amount changes sinusoidally in time if the orbits are circular.

In Fig. 9.5 we show the spectra of a binary system, in which the stars are too close to be observed separately. We can only obtain spectra of both stars together, which show the spectral lines of both stars. We see a time series of spectra, proceeding through one orbital period from the top spectra to the

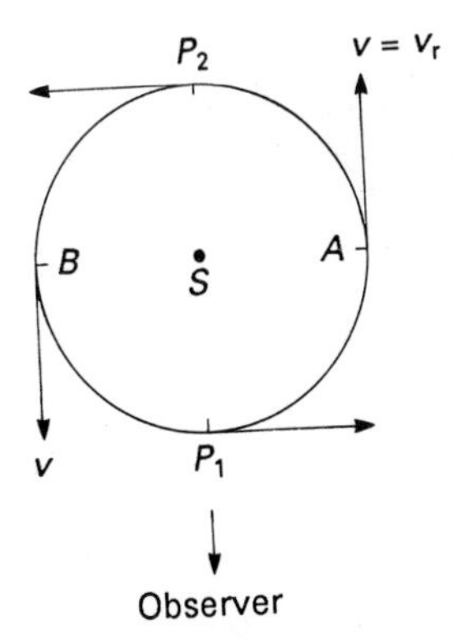

Fig. 9.4. At points *A* and *B* the spectra show Doppler shifts of the lines corresponding to the orbital velocities. At the positions P_1 and P_2 the radial velocities are zero.

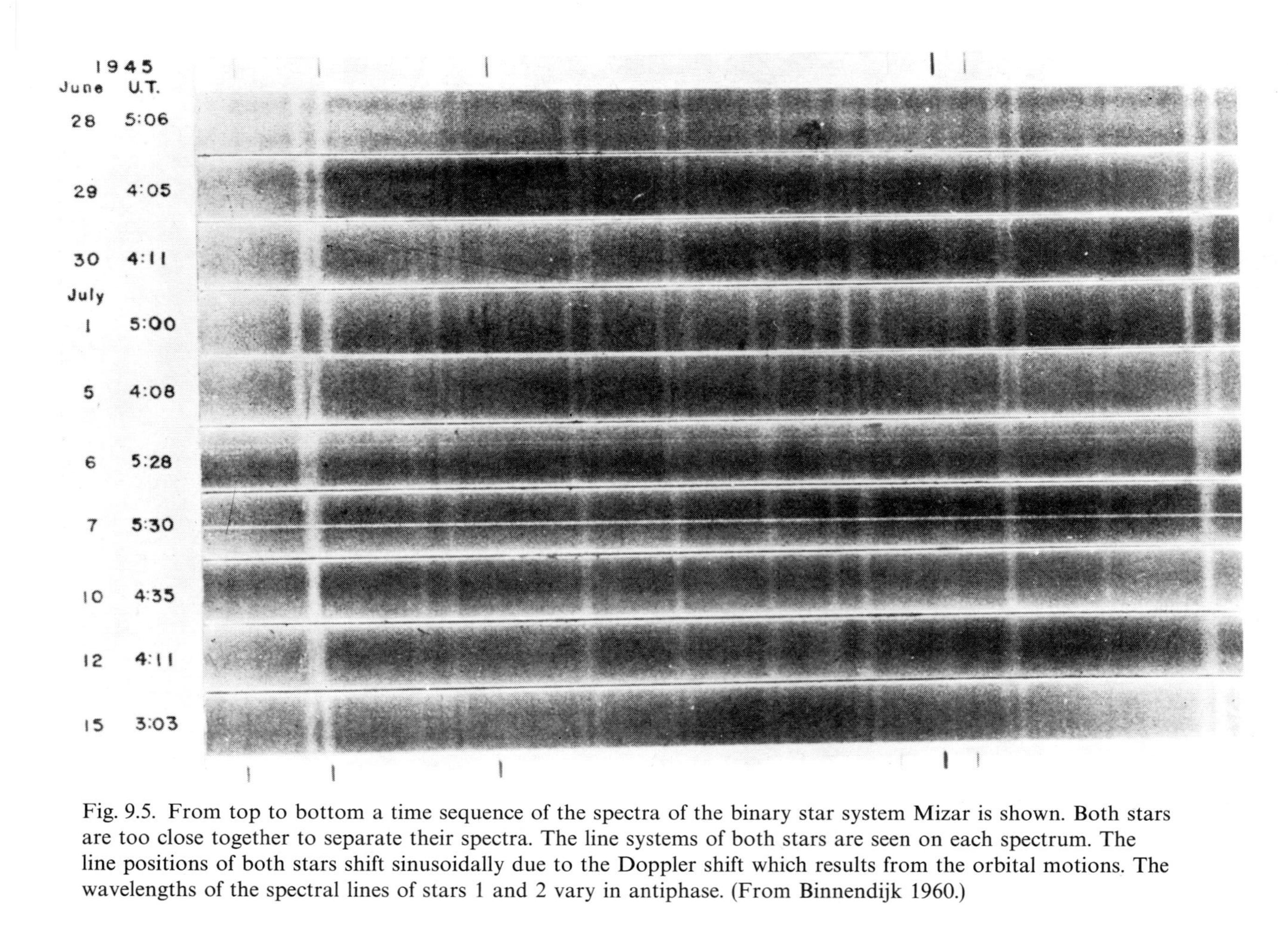

Fig. 9.5. From top to bottom a time sequence of the spectra of the binary star system Mizar is shown. Both stars are too close together to separate their spectra. The line systems of both stars are seen on each spectrum. The line positions of both stars shift sinusoidally due to the Doppler shift which results from the orbital motions. The wavelengths of the spectral lines of stars 1 and 2 vary in antiphase. (From Binnendijk 1960.)

bottom spectra. The approximately sinusoidal change in separation between the lines of the two stars can be seen.

Radial velocity curves as measured for the two stars in the binary system ζ Phoenicis are reproduced in Fig. 9.6B.

Stars which cannot be observed separately but whose binary nature can be seen from the periodic Doppler shifts of their spectral lines are called spectroscopic binaries.

For binaries, for which both spectra can be obtained, and for *circular orbits* with period P we must have

$$v_1 = \frac{2\pi r_1}{P} \quad \text{and} \quad v_2 = \frac{2\pi r_2}{P} \tag{9.11}$$

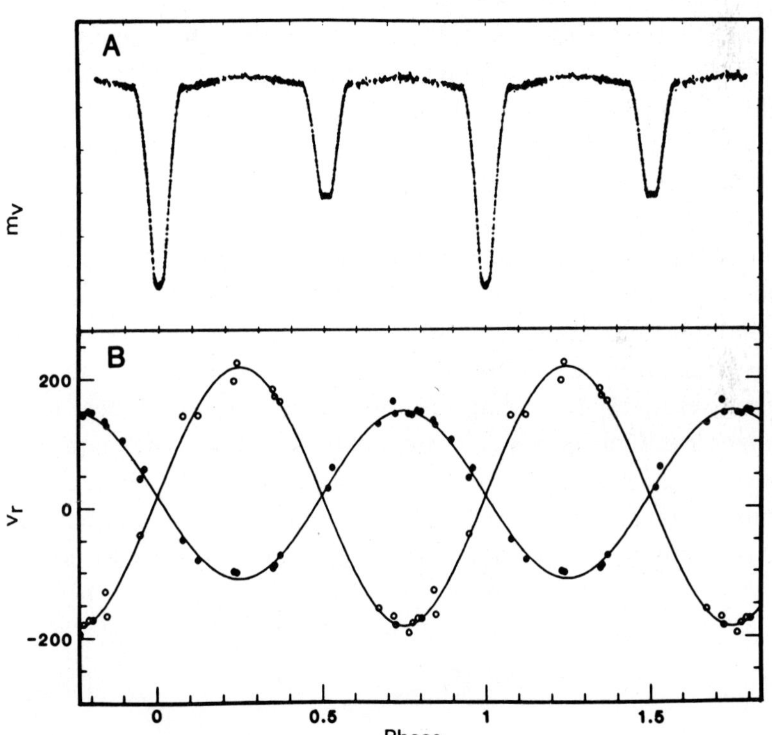

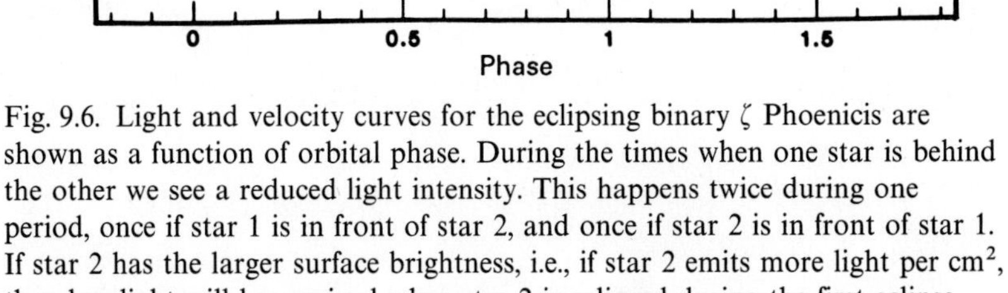

Fig. 9.6. Light and velocity curves for the eclipsing binary ζ Phoenicis are shown as a function of orbital phase. During the times when one star is behind the other we see a reduced light intensity. This happens twice during one period, once if star 1 is in front of star 2, and once if star 2 is in front of star 1. If star 2 has the larger surface brightness, i.e., if star 2 emits more light per cm^2, then less light will be received when star 2 is eclipsed during the first eclipse than will be received when star 1 is eclipsed, i.e., during the second eclipse. The first light minimum is then deeper than the second one. (From Paczinski 1985.)

and

$$\frac{v_1}{v_2} = \frac{r_1}{r_2} = \frac{M_2}{M_1}. \tag{9.12}$$

The periods must of course be the same for both stars. v_1 and v_2 can be measured at maximum angular distance of the two stars. With the measured period, r_1 and r_2 and the sum $r_1 + r_2$ can be determined in cm, which then also gives the sum of the masses according to (9.8). From (9.12) the mass ratio can be obtained. With the mass sum and the mass ratio known it is a simple mathematical problem to determine both masses.

Unfortunately, most stars do not have circular orbits and our line of sight usually is not in the orbital plane.

If the stars are not in circular orbits, their radial velocity curves are in general not sinusoidal, the shape depends on the degree of eccentricity and on the orientation of the major axis of the orbit with respect to the line of sight. It also depends on the tilt of the orbital plane with respect to the line of sight, which is described by the angle i between the normal **n** on the orbital plane and the line of sight.

If the inclination angle i between the normal **n** on the orbital plane and the line of sight is not 90°, which means the line of sight is not in the orbital plane, then all measured radial velocities only measure the component of the velocity projected onto the line of sight. This means all velocities are reduced by the factor $\sin i$. Even for circular orbits the radial velocity measured at the largest distance is then only $V \sin i$. (see Fig. 9.7). Then we cannot determine the masses without knowing $\sin i$. For $i = 90°$ the stars appear to move in a straight line. Only for $\sin i = 0$ are we able to see the true shape of the orbit, but then we cannot measure the velocities. For angles i between 0 and 90° orbits appear as ellipses even if they should be circular. In order to be able to derive semi-major axes and masses we therefore have to determine the

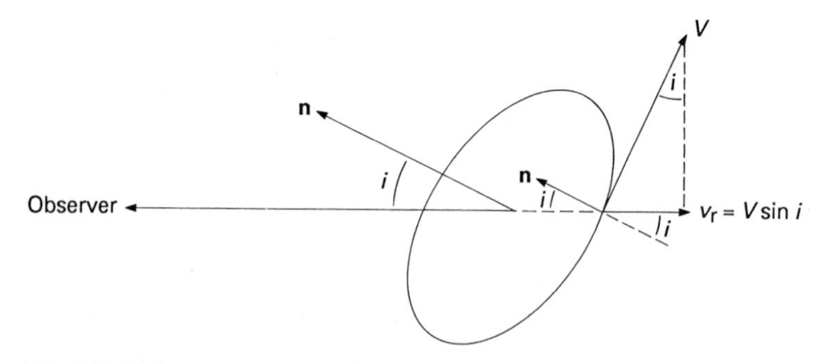

Fig. 9.7. If the normal **n** to the orbital plane is inclined with respect to the line of sight by the angle i, then the radial velocity v_r is $V \sin i$ if V is the true orbital velocity.

orbital parameters like ellipticities, orientation of the ellipse and the inclination angle i of the orbital plane.

9.4 Determination of orbital parameters for binaries

9.4.1 General discussion

Before we discuss in detail the procedure for determining the orbital parameters for binaries let us see how the radial velocity curves change for different orbits. The orbit is determined if we know the ellipticity, the orientation of the ellipse and the inclination angle i between the line of sight and the normal **n** on the orbital plane.

The ellipticity e is defined by

$$e^2 = 1 - \frac{b^2}{a^2} \tag{9.13}$$

where a and b are the semi-major and semi-minor axes respectively. The orientation of the ellipse is described by the angle ω, which is the angle between the minor axis of the ellipse and the projection of the line of sight onto the orbital plane.

In Fig. 9.8 we show radial velocity curves to be observed for a circular orbit, $e = 0.0(a)$, and for $e = 0.5$ for different orientations of the ellipse. For different angles i all the velocities would be multiplied by the factor $\sin i$ but the shape would remain the same. We cannot therefore determine the angle i from the shape of the velocity curve. The ellipticity can however be determined from the shape of the radial velocity curve. If $e \neq 0$ the curve is not sinusoidal because the stars have larger velocities when they are closer together. In Fig. 9.8(b) we see the high peaks of the radial velocity when the star moves through point b, the perihelion of its orbit. When it moves towards us at point d it is far away from the other star and therefore has a much lower velocity. We measure a broad, flat valley for the negative velocities. A very asymmetric velocity curve is obtained if $\omega = 45°$ as seen in Fig. 9.8(c), but we still see larger positive velocities than negative ones. If $\omega = 90°$ (Fig. 9.8(d)) the position of the ellipse is symmetric with respect to the line of sight. The largest positive and negative velocities have therefore equal absolute values, but it takes the star a much longer time to get from the largest negative velocity to the largest positive velocity, because the distance is larger and the speed is smaller when going from d to b than when going from b to d. From Fig. 9.8 we can easily see that we can determine e and ω from the deviations of the radial velocity curve from a sinusoidal curve.

For the details of the mathematical procedure see for instance Binnendijk 1960.

9.4.2 *Determination of i for visual binaries*

From the study of the radial velocities we have no means of determining the inclination angle i between the line of sight and the normal **n** to the orbital plane. Since the shape of the radial velocity curve is not changed by a change in i we have no way to determine i if we have only radial velocity curves. If we are however dealing with a visual binary for which we can measure the shape of the orbits we can determine i. We know that for binaries the second star must be in the focal point of the orbit for the first star. Suppose the stars have circular orbits which appear as ellipses because i is not 0. For a circular orbit the secondary star will therefore appear in the

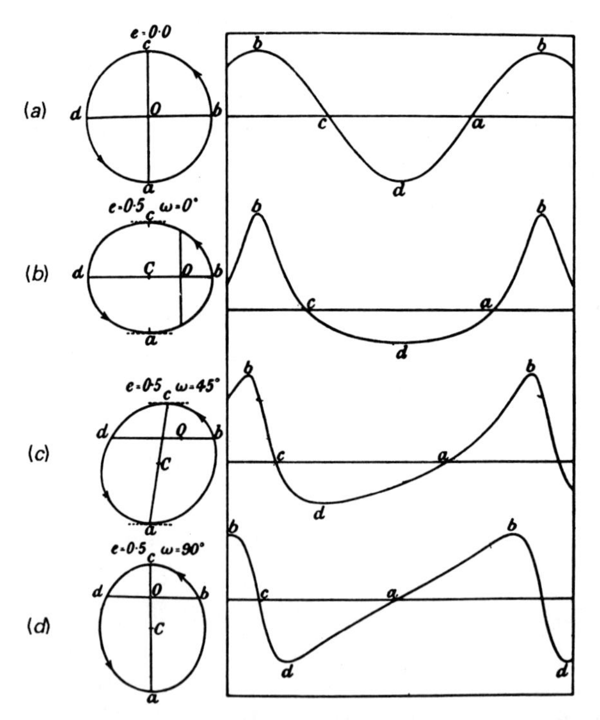

Fig. 9.8. Plotted are radial velocity curves to be seen for stellar orbits with zero ellipticities (a) and with an ellipticity $e = 0.5$ for $\omega = 0$ (b), and for the same ellipticity but $\omega = 45°$ (c), and also for $\omega = 90°$ (d). $e \neq 0$ generally introduces a difference between maximum positive and negative velocities (see (c) and (d)) except in the case when $\omega = 90°$. In this case the time difference between points b and d on the one side and between points d and b on the other side (d) gives a measure of the ellipticity. (From Becker 1950.)

middle of the major axis and not in the focal point (see Fig. 9.9). We know immediately that the true orbit must be a circle. From the apparent ratio of the minor to the major axis we can in this special case, determine how large i is. If the second star appears off center but not in the focal point we also know that i is not 90°, but that the true orbit is an ellipse. We can figure out how much we have to tilt the observed projected ellipse in order to fit the second star onto the focal point of the orbital ellipse.

For the procedure and details of the orbit determinations see for instance Binnendijk 1960.

9.4.3 Distance to visual binaries

In Section 9.3 we saw that we can determine the orbital radii in cm if we can determine the true orbital velocity by measuring the radial velocities, determining the orbital parameters, and determining $\sin i$. If we know the true orbital velocities then the orbital radii and the velocities are related as given by (9.11) for the case of a circular orbit. From the two Equations (9.11) we can determine r_1 and r_2 and thereby $r_1 + r_2$. Observing the orbit of one star around the other permits us to measure the angular semi-major axis $(r_1 + r_2)/d = \alpha$, where d is the distance to the binary system. Once we have determined r_1 and r_2 in cm we can also find d in cm (or in pc) from

$$d = \frac{r_1 + r_2}{\alpha}. \tag{9.14}$$

For visual binaries we can therefore determine the distance even if we cannot measure trigonometric parallaxes.*

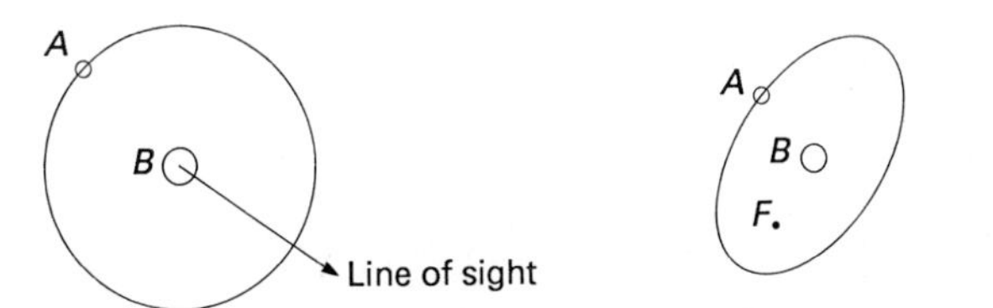

Fig. 9.9. For a circular orbit of star A around star B whose normal **n** is inclined at an angle i with respect to the line of sight, the apparent orbit projected against the background sky is an ellipse with the second star B in the middle of the major axis, instead of in the focal point F.

* In practice there are only a few systems which are close enough for the two stars to be seen separately while the periods are not too long to be determined in a (wo)man's lifetime and for which the radial velocities are large enough to be accurately measurable.

9.4.4 Eclipsing binaries

a. *Mass determination.* Another group of binaries for which i can be determined are the eclipsing binaries. For these stars the line of sight is almost in the orbital plane such that at some time the secondary star is in front of the primary (see Fig. 9.10). Half a period later the primary star will be in front of the secondary. If along the line of sight one star gets in front of the other, we can be fairly certain that $\sin i$ has to be close to 1 otherwise the 'front' star would remain above or below the 'back' star, except if the stars are very close (see Fig. 9.11). By far the most binaries are separated far enough that eclipses will only occur if $i > 75°$ which means $\sin i \geqslant 0.96$ and $\sin^3 i \geqslant 0.89$. This means that if any sign of an eclipse is seen $\sin^3 i$ cannot differ by more than 10% from 1. Assuming $\sin^3 i = 1$ will at most cause a 10%

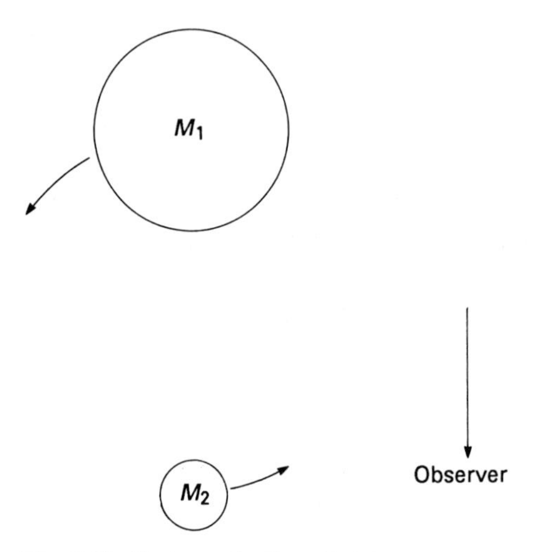

Fig. 9.10. If along the line of sight one star is in front of the other star system the total amount of light of the system is reduced. We see the eclipse of the star.

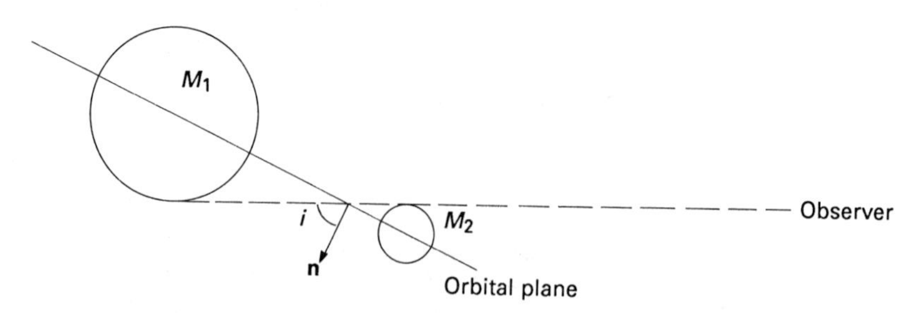

Fig. 9.11. If the line of sight is inclined too much to the orbital plane then star M_2 is not in front of star M_1 along the line of sight. The angle i has to be very close to 90° for an eclipse.

error in the masses (see equation 9.26). With refined technique we can even determine $\sin i$ for partial eclipses. We can usually assume $\sin i = 1$ for eclipsing binaries.

Such eclipses can be recognized by the decreasing amount of light received during the phases of occultation. In Fig. 9.6 we showed the light curve for the eclipsing binary ζ Phoenicis. In Fig. 9.12 we reproduce schematic light curves for binaries with different ratios of the size of their radii and with different surface brightnesses. In Fig. 9.13 we relate the geometrical situation to the observed light curve as a function of time. From Fig. 9.14 we can see, that for a circular orbit of star A around star B the ratio of the total length of the eclipse time t_e to the length of the period P is given by

$$t_e/P = \frac{2R_A + 2R_B}{2\pi r}, \tag{9.15}$$

where R_A and R_B are the radii of stars A and B respectively, and r is the orbital radius of star A around star B. Equation (9.15) assumes that star B is much more massive than star A such that the velocity of star B can be neglected. It also assumes $r \gg R_A$ and R_B.

If $t_e/P \ll 1$ the stellar radii must be much smaller than the orbital radius and the stars cannot be very close. Eclipses must then mean that $i \approx 90°$ and

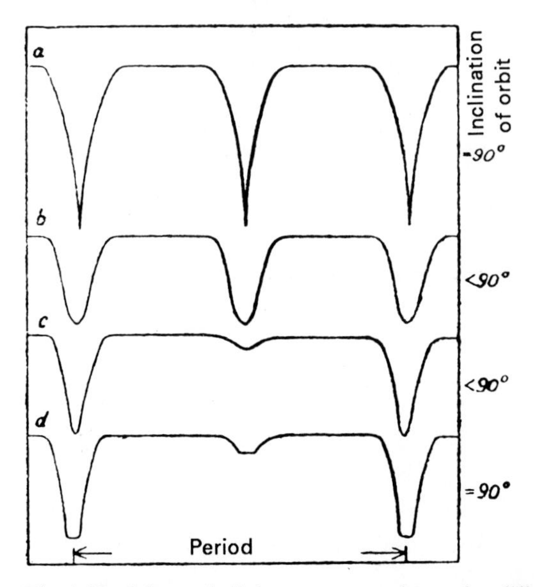

Fig. 9.12. Schematic light curves are shown for different ratios of stellar radii and different surface brightnesses of the eclipsing stars. Curves (*a*) and (*b*) show the light curves for eclipsing binaries for which both stars have the same radius and the same surface brightnesses. Curves (*c*) and (*d*) apply to binaries with different radii and surface brightnesses. From Becker (1950).

$\sin i \approx 1$. We can get even more accurate information about i if we study the shape of the light curve in detail. The decrease of the light intensity will be less steep if the secondary star does not pass over the equator of the other star (see Fig. 9.13(*b*)).

For such eclipsing binaries we then also know the inclination angle i. We can therefore determine the masses of both components as described in Sections 9.4.1 and 9.4.4.

b. *Radius determination.* Eclipsing binaries are also very useful for radius determinations, especially if we see a total eclipse.

Fig. 9.12 shows light curves which are obtained for stars with different radii and different surface brightnesses. Since these light curves look very different we can use the light curves to determine the properties of the stars. For simplicity we again assume circular orbits. Generally orbits are elliptic

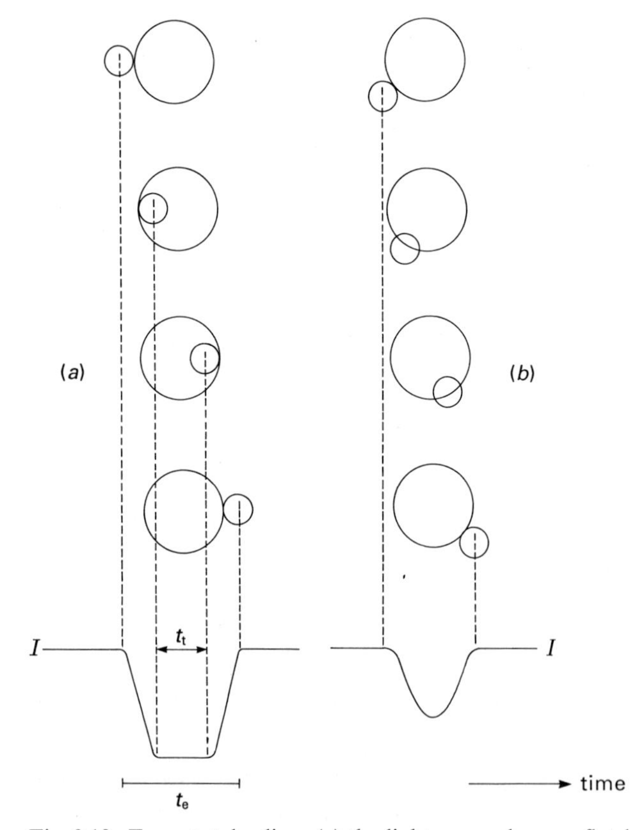

Fig. 9.13. For a total eclipse (*a*) the light curve shows a flat bottom (except if the radii are equal), while for a partial eclipse (*b*) the light curve does not have a flat bottom.

but the method does not change, only the mathematics gets more complicated.

If radial velocities for the stars are measured we can determine the radii of both stars by means of (9.15). The maximum radial velocity of star A is

$$V_{\mathrm{r(max)}} = 2\pi r_A/P. \tag{9.16}$$

We also know that for a total eclipse the duration for the flat part of the light curve t_t, the time during which star A is totally in front of star B, is related to the length of the period P by

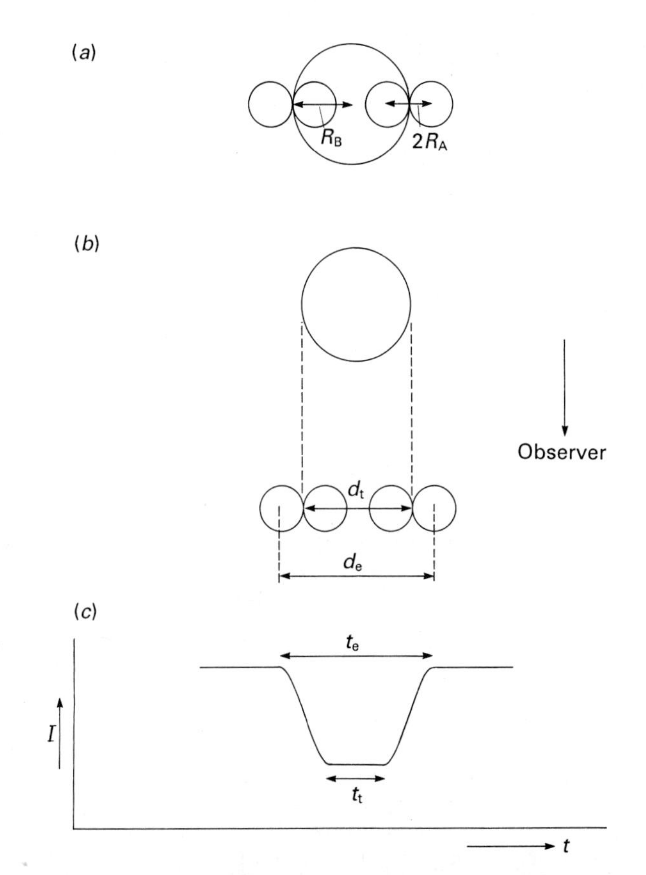

Fig. 9.14. Figure (*a*) shows the positions of the eclipsing star at first, second, third, and fourth contact as seen from an observer above the plane of the paper. Figure (*b*) shows the geometry of the eclipse for an observer whose line of sight is in the orbital plane (which is the plane of the paper), not drawn to scale. For a large distance between stars A and B the path of star B can be approximated by a straight line. Figure (*c*) shows the light curve as a function of time for the duration of the eclipse. During the time t_e star A travels a distance $d_e = 2R_A + 2R_B$. During the time t_t star A travels the distance $d_t = 2R_B - 2R_A$.

$$t_t/P = (2R_B - 2R_A)/2\pi r_A. \tag{9.17}$$

if again the mass of star B is much larger than the mass of star A. Equation (9.16) determines the orbital radius r_A. From (9.15) and (9.17) we obtain

$$(t_e - t_t)/P = \frac{4R_A}{2\pi r_A} \tag{9.18}$$

and

$$(t_e + t_t)/P = \frac{4R_B}{2\pi r_A}. \tag{9.19}$$

These equations determine the radii of both stars.

If stars A and B have comparable masses we have to consider the motions of both stars. The eclipse times will then be shorter, because the relative velocities of the two stars will be larger, namely $V_A + V_B$, with $V_A = 2\pi r_A/P$ and $V_B = 2\pi r_B/P$.

If the stars have elliptical orbits we have to know in which part of the ellipse the occultation occurs, such that we can relate the maximum measured radial velocity to the relative velocity at the time of the eclipse. The mathematics becomes more involved but the principle of determining the radii of both stars remains the same.

c. *Surface brightness of eclipsing binaries.* From the light curves of eclipsing binaries we can also determine the ratio of the surface brightnesses of the stars.

From Fig. 9.14 we see that maximum light is given by

$$I_{\max} = (\pi R_B^2 \cdot F_B + \pi R_A^2 \cdot F_A)\text{const.} \tag{9.20}$$

where F_A and F_B are the amounts of radiation emitted per cm^2 of the stellar surface of star A and star B respectively.

The constant is determined by the distance and the sensitivity of the receiving instrument as well as the transmission of the Earth's atmosphere. During the first eclipse the minimum intensity I_1 is given by (see Fig. 9.15)

$$I_1 = [(\pi R_B^2 - \pi R_A^2) \cdot F_B + \pi R_A^2 \cdot F_A]\text{const.}. \tag{9.21}$$

During the second eclipse the minimum intensity I_2 is given by

$$I_2 = (\pi R_B^2 \cdot F_B)\text{const.}. \tag{9.22}$$

From this we derive

$$I_{\max} - I_2 = (\pi R_A^2 \cdot F_A)\text{const.} \tag{9.23}$$

and

$$I_{\max} - I_1 = (\pi R_A^2 \cdot F_B)\text{const.}. \tag{9.24}$$

Knowing R_B and R_A from our previous discussion we can then determine F_A and F_B from (9.23) and (9.24), except that we have to know the constant, which means we have to know the distance, the transmission of the Earth's atmosphere, the sensitivity of the instrument, etc. So in general we can only determine ratios namely

$$\frac{I_{\max} - I_2}{I_{\max} - I_1} = \frac{F_A}{F_B}. \tag{9.25}$$

If we know the flux for one star, we can determine the flux of the other star. Since the radiative fluxes F_A and F_B determine the effective temperatures of the stars we can also determine the ratio of the effective temperatures of the two components of an eclipsing binary system.

9.4.5 *Spectroscopic binaries*

Most binaries are spectroscopic binaries, which means we can see their binary nature from the periodic line shifts due to the Doppler effect. Very often we see only lines from one star because the companion is too faint to be recognized in the combined spectrum. If we see both spectra we can determine the mass ratio of the two stars according to (9.2). It does not cause any problems that we can only measure $v_1 \sin i$ and $v_2 \sin i$, without knowing what $\sin i$ is. In the ratio the $\sin i$ factor cancels out. But if we want to determine the mass sum we are in trouble. We know that $r_1 = v_1 P/(2\pi)$ and similarly $r_2 = v_2 P/(2\pi)$. Since we only know $v_1 \sin i$ and $v_2 \sin i$ we also can

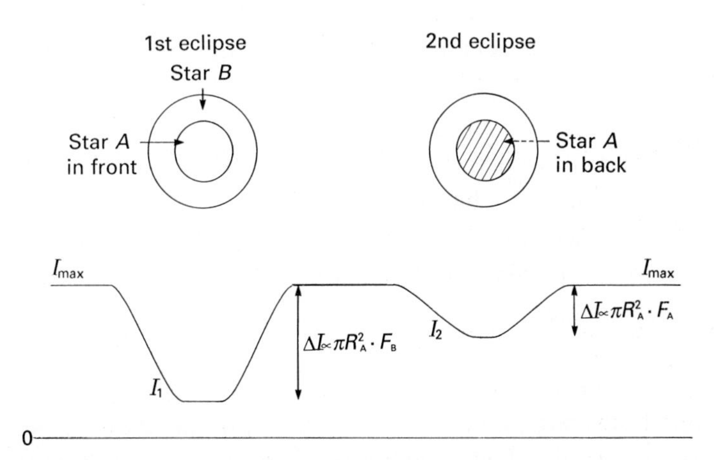

Fig. 9.15. During the first eclipse a fraction of the surface of star B is occulted. During the second eclipse star A is completely covered up. The depths of the light minima depend on the surface fluxes of star A and star B.

Table 9.1 *Masses and radii for some visual binaries according to D. Popper (1980)*

Star	S_p	$M/M_\odot$	$R/R_\odot$	π''
α CMaA	A1 V	2.20	1.68	0.377
α CMi	F5IV-V	1.77	2.06	0.287
ξ Her A	G0IV	1.25	2.24	0.104
ξ Her B	K0V	0.70	0.79	0.104
α Cen	G2V	1.14	1.27	0.743
γ Vir	F0V	1.08	1.35	0.094
η Cas A	G0V	0.91	0.98	0.172
η Cas B	M0V	0.56	0.59	0.172
ξ Boo	G8V	0.90	0.77	0.148

only determine $r_1 \sin i$ and $r_2 \sin i$. According to (9.8) we therefore can determine only

$$(M_1 + M_2)\sin^3 i = \frac{(r_1 + r_2)^3 \sin^3 i}{P^2} \frac{4\pi^2}{G}. \tag{9.26}$$

The mass sum remains uncertain by the factor $\sin^3 i$, and therefore both masses remain uncertain by the same factor.

9.5 Data for stellar masses, radii, and effective temperatures

In Table 9.1 we list stellar masses as determined from the study of visual binaries according to D. Popper (1980). In the same table we also list stellar radii determined for the same binary systems.

In Table 9.2 we have given average values of masses, radii and effective temperatures for main sequence stars of different B − V colors, corrected for interstellar extinction (see Chapter 19).

In Fig. 9.16 we have plotted a color magnitude diagram with the position of the average main sequence drawn in. On this average main sequence we have also given values for the mass, radius, and T_{eff} for stars at that particular position on the main sequence. Also given are spectral types, which will be discussed in the next chapter.

At the top of the main sequence we find the most massive stars, which are also the largest ones and the hottest ones. It becomes obvious that the main sequence is indeed a one parameter sequence: If we know the B − V color for a star we also know its absolute magnitude, its mass and radius and its

Table 9.2 *Masses and radii for main sequence stars according to Schmidt-Kaler (1982) and Popper (1980)*

Spectral[1] Type	B − V	$R/R_\odot$	$M/M_\odot$
O8	−0.32	10	23
B0	−0.30	7.5	16
B2	−0.24	5.4	9
B4	−0.18	4.2	6
B6	−0.15	3.4	4.5
B8	−0.11	2.9	3.8
A0	−0.01	2.4	2.8
A5	+0.15	1.6	2.0
F0	+0.30	1.4	1.6
F5	+0.44	1.3	1.3
G0	+0.58	1.1	1.1
G5	+0.68	0.95	0.93
K0	+0.81	0.85	0.82
K5	+1.15	0.72	0.65
M0	+1.40	0.60	0.65
M5	+1.62	0.27	0.20

1. Spectral types will be discussed in Chapter 10.

effective temperature. In Volume 3 we will try to explain why most of the stars form a one-dimensional sequence.

9.6 The mass–luminosity relation

As we saw in the previous section there is a relation between mass and the absolute visual magnitude for the stars on the main sequence, the more massive the star is, the brighter it is in the visual. We also saw in Section 6.3 that for stars hotter than 8000 K the bolometric corrections increase and therefore the bolometric luminosities also increase for increasing mass of the stars. For stars cooler than 7000 K the bolometric corrections increase for stars with smaller visual brightness, but we still find decreasing bolometric brightnesses for decreasing masses of the stars. On the main sequence we therefore find a simple relation between mass and luminosity: the larger the mass the larger the luminosity. This may not be surprising since we have just emphasized that the main sequence is indeed a one-dimensional sequence and therefore all the stellar parameters are related in a unique way. What is surprising, however, is the fact that the mass–

luminosity relation also holds approximately for the giants and supergiants. We find several of them in binaries for which we can determine the masses. In Fig. 9.17 we show the mass–luminosity relation as seen for all well-determined masses in binaries. The exceptions are the red giants which for a given luminosity have generally smaller masses. In Volume 3 we will explain this result which at first sight appears very surprising.

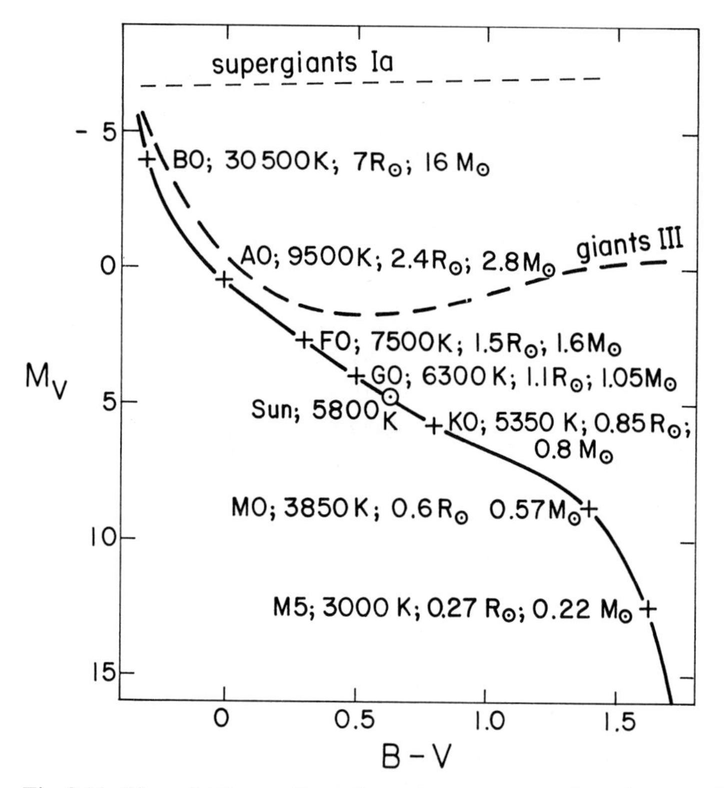

Fig. 9.16. The solid line outlines the main sequence in the color magnitude diagram. At the different points of the main sequence the masses, radii, and effective temperatures of the stars are given.

The dashed line shows the position of the giant sequence, and the dotted line the approximate positions of the brightest supergiants, called luminosity class Ia.

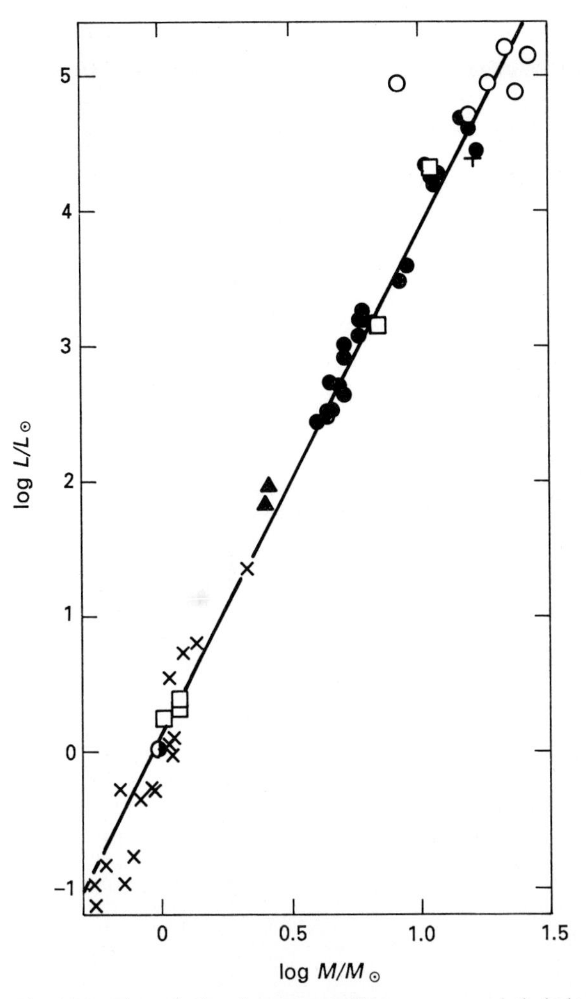

Fig. 9.17. The relation between stellar masses and their luminosities is shown for binaries with well determined masses, according to Popper (1980). The different symbols refer to different kinds of binaries. The open circles refer to O type binaries, the filled circles to O, B pairs, the × to visual binaries, and the triangles to giants. The open squares refer to resolved spectroscopic binaries.

10

Spectral classification

10.1 The spectral sequence

So far we have talked only about global properties of the stars and about their brightnesses in broad wavelength bands. We get, of course, more information when we reduce the widths of the wavelength bands in which we study the radiative energy emitted by the stars. If we reduce the bandwidth to the order of a few Å or even a fraction of an Å, and cover all wavelengths, we talk of stellar spectra. If we compare spectra of different stars, we see that there are many different kinds of spectra. Most of them can be ordered in a continuous sequence of spectra, the so-called spectral sequence. In Fig. 10.1 we show the sequence as it is used now, and as it was established finally by Morgan, Keenan and Kellman (1943).

If we plot the energy distribution in a spectrum as a function of wavelength, we get plots as shown in Figs 10.2. Basically, we see a continuous energy distribution (Fig. 10.2(*a*)), but there are many wavelengths for which the energy is reduced by varying amounts, the so-called spectral lines, see Fig. 10.2(*b*). If we look at these wavelengths in a spectrum, they look dark because there is little energy at these wavelengths. Such dark lines are called absorption lines. There are also some spectra which have bright lines, i.e., for which there is more energy at these wavelengths. Such lines are called emission lines. Most stellar spectra show absorption lines. This is quite different from the situation which is known from the chemistry laboratory, where emission line spectra are used to identify chemical elements. Kirchhoff and Bunsen discovered in 1859 that an absorption line spectrum is seen if in the laboratory we place a cold gas in front of a hot light source (see Fig. 10.3). The fact that we see absorption lines in the stellar spectra thus tells us that the gas at the stellar surface is cooler than the gas in the somewhat deeper layers (see Fig. 10.4). In Volume 2 we will understand better when emission lines and when absorption lines are formed.

The spectral classification is done according to the lines and the line

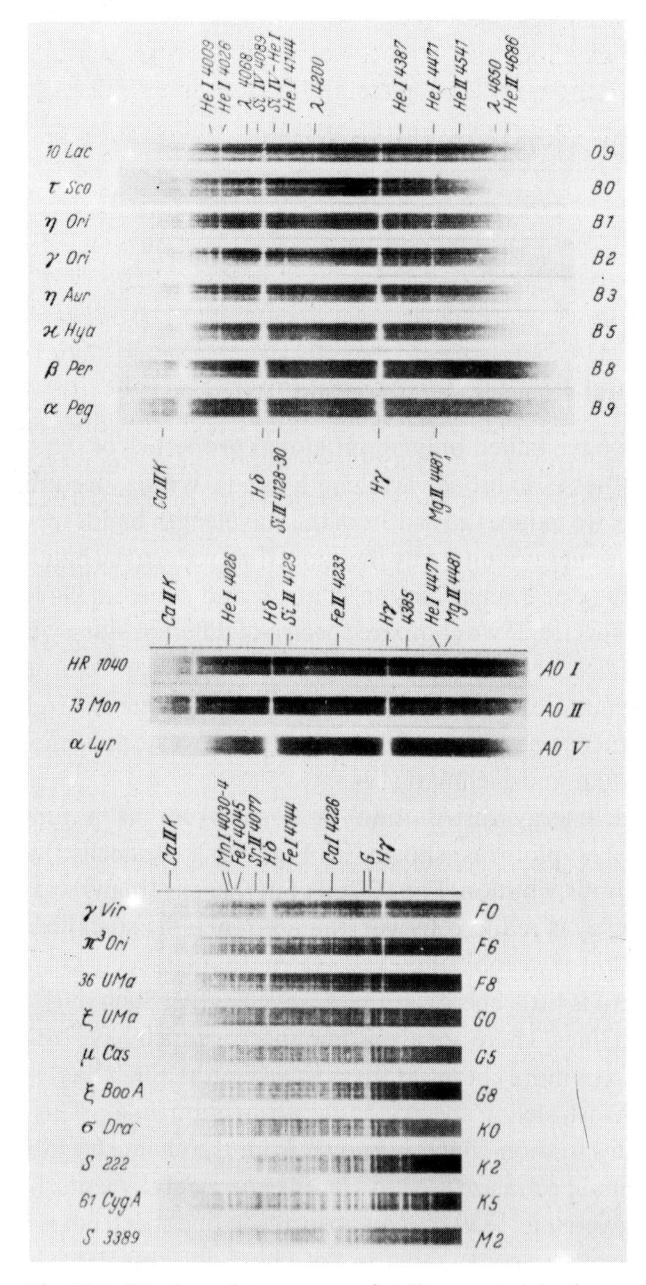

Fig. 10.1. We show the sequence of stellar spectra. At the top of the figure the O type spectrum of the star 10 Lac is reproduced. At the bottom we see an N type spectrum as observed for the star S 3389. All the stars for which the spectra shown have been observed are indicated on the left side. For a description of the spectral types see the text. Shown are negatives. The white lines are wavelengths in which no or very little energy is received, i.e. they are absorption lines.

strengths seen in the spectra. These lines are also called the Fraunhofer lines after the optician who first discovered them in the solar spectrum. Originally when astronomers did not then understand the origin and the meaning of these spectral lines, they realized that there were a number of strong lines which were strong in almost all spectra, so they put the spectra in a sequence according to the strengths of these lines. We now know that these lines are due to absorption by the hydrogen atoms in the stars. The early astronomers called the spectra with the strongest hydrogen lines, spectral type A, those with somewhat weaker hydrogen lines were called spectral type B, and went through the alphabet for weaker and weaker hydrogen lines, down to spectral types M, N, O, R, S. It turned out, however, that within this sequence the line strengths of the other lines varied quite irregularly, also the B − V colors varied irregularly along the sequence. It was much better to put the spectra into a sequence which also took into account the changes of the other

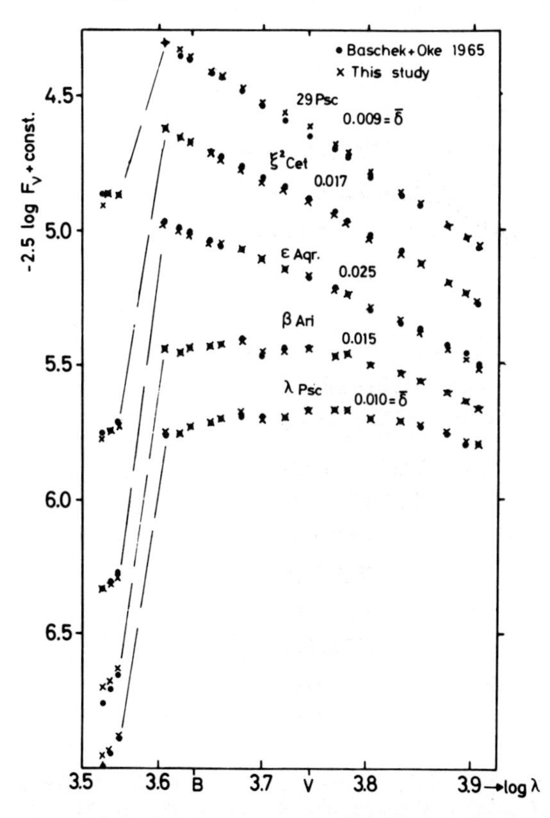

Fig. 10.2(*a*). Continuous energy distributions for A stars are shown as a function of wavelength for the wavelength interval 3000 Å $< \lambda <$ 9000 Å. $\bar{\delta}$ are the standard deviations for the measurements. The energy distributions of A stars are strongly wavelength dependent, especially around 3800 Å where the Balmer discontinuity is seen. (From Böhm-Vitense & Johnson 1977.)

lines. After the reordering had been done, it appeared that the new spectral sequence was also a sequence according to the colors of the stars. As usual, in the meantime the astronomers had become accustomed to the old spectral types and did not want to change the spectral type designation for their stars.

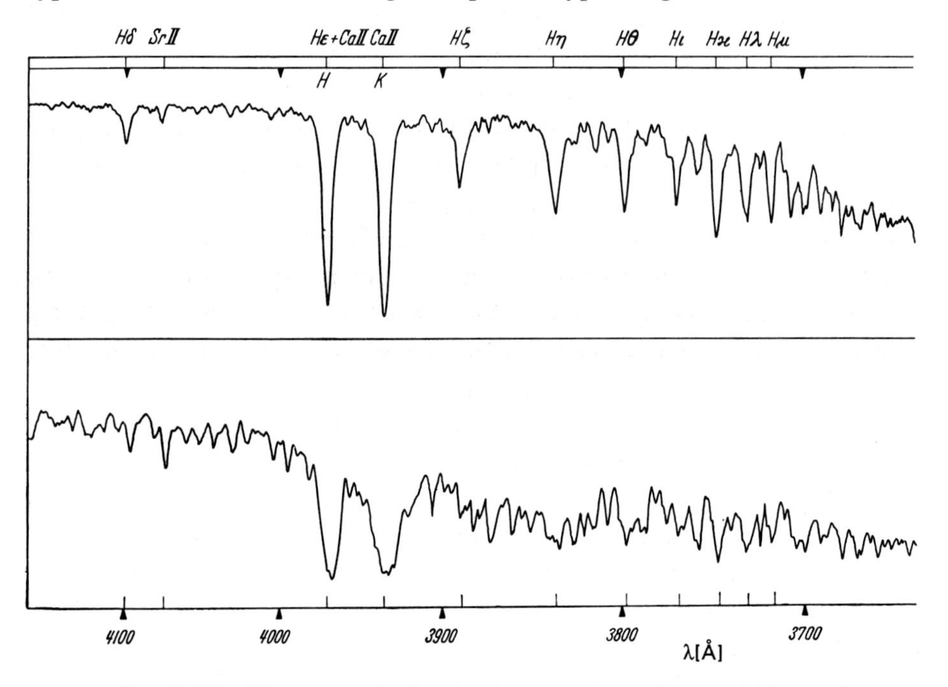

Fig. 10.2(*b*). The energy distributions in two spectra of the variable star δ Cephei are shown as a function of wavelength. The upper spectrum refers to the phase of maximum light and maximum temperature when the star has a spectral type F5 Ib. The hydrogen Balmer line series is clearly visible. The lower spectrum refers to the phase of minimum light and minimum temperature, when the star has a spectral type G2Ib. At this time the star is so cool that the hydrogen lines are barely visible for this low dispersion. The Ca^+ lines are very strong and many other lines of metallic ions and atoms are seen. (From Ledoux & Walraven 1958.)

Fig. 10.3. Cold Na vapor in front of a light source emitting a continuous spectrum absorbs light at the wavelengths of 5889 and 5896 Å, giving dark absorption lines in the spectrum. A hot light source with Na vapor in it would show emission lines, as seen in the Bunsen burner.

So now we still have most of the original spectral types, but they do not appear in alphabetical order in the new sequence. The A stars with the strongest hydrogen lines are now in the middle of the sequence. The sequence now goes O, B, A, F, G, K, M. At the cool end, the classification becomes more complicated. In addition to the M stars, we also have *N* stars and S stars, depending on which molecular bands are seen.

The O, B, and A types are called the early types, just because A and B are 'early' in the alphabet. The term 'early' has nothing to do with the age of the stars. Correspondingly, the G, K, M, N, and S stars are called late type stars because these letters are later in the alphabet.

The sequence now starts with spectral type O, which indicates that these stars have rather weak hydrogen lines, but they are the bluest stars. In the spectra of these stars we can recognize lines which we know from laboratory experiments to be due to the helim atom and the helium ion, i.e., a helium atom which has lost one electron. The O stars are the only stars in whose spectra the lines of ionized helium can be seen. It takes a lot of energy, 24 electron volts (ev), to separate the electron from the helium atom. Only in very hot stars do the particles have enough energy to ionize the helium atom.

The next spectral type in the new sequence is type B. As the letter indicates, these spectra show rather strong hydrogen lines but they also show lines of the helium atom, thereby indicating that they should be placed close to the O stars. Most other lines in the spectra of the B stars are very weak.

The next type in the sequence are the A stars with the strongest hydrogen lines. They do not show helium lines, or at least only extremely weak ones which can only be seen if very high spectral resolution is used, which means if the energy distribution is measured in very narrow wavelength bands. The A

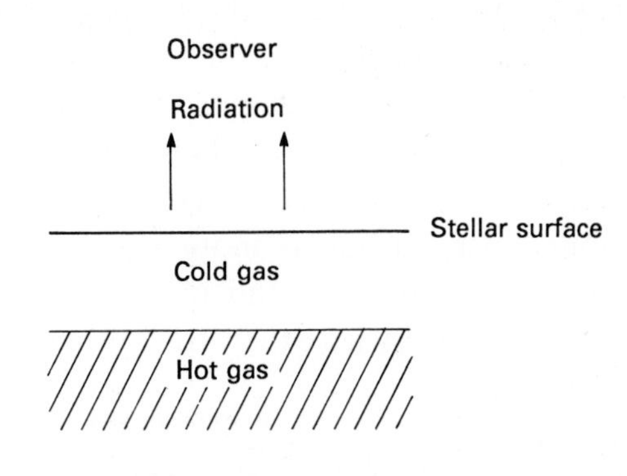

Fig. 10.4. The stellar radiation comes from a hot interior and passes through the cooler surface layers. The spectrum which we observe therefore shows absorption lines.

stars also begin to show a number of other spectral lines which, as we know from laboratory experiments, are due to heavier elements like iron (Fe), chromium (Cr), and many other elements. The lines seen are due mainly to the ions of these elements, which means again that these atoms have lost one of their electrons. It takes only about 8 eV to ionize these elements.

For the F stars, which follow the A stars, these lines from the ions of the heavier elements become stronger, and we also start to see lines which are due to the neutral particles, the atoms.

Astronomers are usually somewhat sloppy and call all the elements whose atoms are heavier than helium the 'metals.' It is clear that, for instance, carbon or oxygen are not metals, but in the astronomer's jargon these elements are often included when the word 'metal' is used. We can then say that for F stars the metallic lines become stronger. This is even more true for the G stars in whose spectra some metallic lines and the hydrogen lines are of comparable strengths. The solar spectrum is of type G. For stars still later in the sequence, the hydrogen lines become still weaker and the metallic lines stronger.

In the K stars we see mainly lines of neutral atoms as the lines of the metallic ions decrease in strength. For the M stars, another type of spectrum becomes visible. We see groups of many lines very close together. These types of spectra are formed by molecules. For stars with M type spectra, also called the M stars, we know that molecules must be present. Indeed, we know that some of these molecular bands, as the line groups are called, are due to molecules CN, CH, and TiO, i.e. cyanogen, carbon hydride and titanium oxide.

Other stars with molecular bands are the R, N, and S stars. They differ from the M star spectra by the kind and strengths of the molecular bands. The S stars do not show TiO bands but instead show ZrO bands. The N and R stars are now usually called C stars, where the C stands for carbon. They show strong bands of various molecules containing a carbon atom, while no molecules with oxygen are seen. These are apparently stars which have a larger than average carbon abundance, carbon is more abundant than oxygen. Under those circumstances the oxygen is all bound in the CO molecule, whose bands are in the ultraviolet where they are not observable for cool stars, because there is so little flux at these short wavelengths.

The C stars are again subdivided into classes C1, C2, C3, etc with increasing numbers for increasing strength of the carbon features in their spectra.

Dividing all the different spectra into just these seven or ten main types is, of course, only a very coarse division which turned out to be much too coarse for modern astronomical research. Each spectral type is, therefore, now subdivided into ten subclasses, in principle, though some of the subclasses are not really in use. The subclasses have numbers. We now have A0, A1, A2,... A9. A9 is then followed by F0, etc. For some spectral classes, even this subdivision did not turn out to be fine enough, so we also have a spectral class B0.5, which means the spectrum is between B0 and B1. We have to get accustomed to the fact that astronomy is a very old science and that everything is built on tradition even though for modern astrophysics many procedures do not make much sense anymore. It would be very difficult to give up old traditions because the old observations and catalogues still have to be used and are very important for many studies.

For the new spectral sequence, the hydrogen lines now have their maximum somewhere in the middle of the sequence and decrease on both sides as shown schematically in Fig. 10.5.

10.2 Luminosity classification

We started out trying to find spectral criteria which would tell us whether a given star is a main sequence star or whether it might be a giant, also called luminosity class III star, or a supergiant, also called luminosity class I star, or perhaps a white dwarf.

For stars in clusters and for very nearby stars, we know from the color magnitude diagram which are main sequence stars and which are giants or supergiants. We can therefore look at the spectra of these stars and see in which way the spectra of the giants and supergiants differ from those of main sequence stars. For blue stars it is obvious that the hydrogen lines become more narrow for luminous stars (see Fig. 10.6). For stars of later spectral

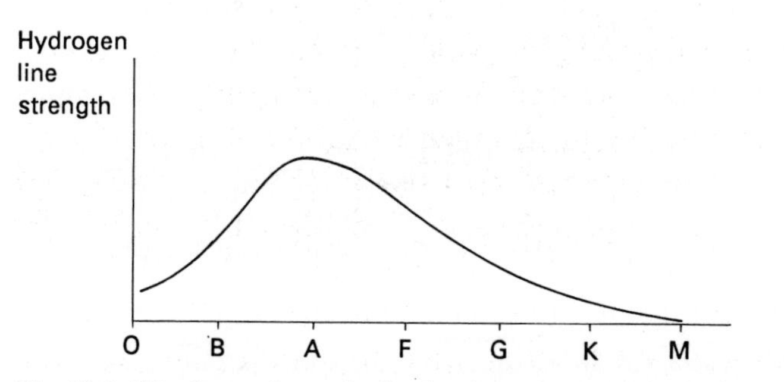

Fig. 10.5. We show schematically the dependence of the hydrogen line strength on the spectral type. When going from the O type stars to later spectral types the hydrogen line strength first increases until we get to the spectral type AO, for later spectral types it decreases.

types this is no longer the case, but other criteria can be found, as demonstrated in Fig. 10.7 in which we show the luminosity criteria for the spectral type F8. Generally the metallic lines increase in strength for supergiants while the hydrogen line strengths do not change for higher luminosity stars of spectral types G or later. The spectral types are determined mainly by ratios of line strengths for metallic lines rather than by the line strengths themselves, because these do change with luminosity.

Once we have learned how to recognize giants and supergiants by means of their spectra, we know whether a given star is a main sequence star or not and whether we can determine its distance by means of spectroscopic parallaxes.

Giants and supergiants have a much larger spread in intrinsic brightnesses. Generally we cannot determine distances of supergiants by means of spectroscopic parallaxes, only rough estimates are possible.

In order to obtain a somewhat better luminosity classification the supergiants are now subdivided into classes Ia and Ib and sometimes also Iab. The Ia supergiants are the brightest ones, then follow the Iab, and the Ib are the faintest supergiants.

The giants are also subdivided into the bright giants, or luminosity class II

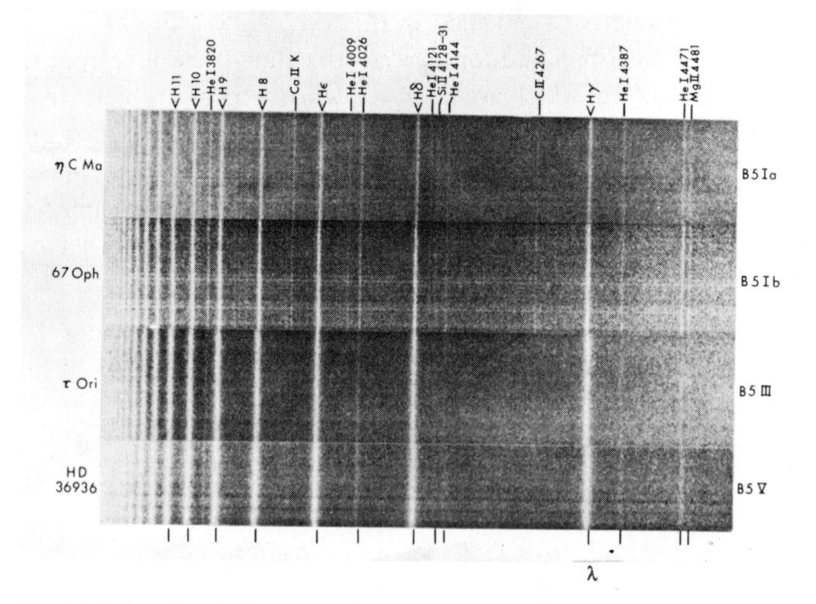

Fig. 10.6. A series of B5 star spectra for stars with different luminosities is shown. The hydrogen lines become narrower and apparently fainter for the more luminous stars. (Luminosity class I means supergiants, II means bright giants, III are giants, IV are subgiants, and luminosity class V are the main sequence stars.) (From Morgan, Abt & Tapscott 1978.)

stars, the 'normal' giants with luminosity class III, and the subgiants with luminosity class IV. The luminosity classification is generally uncertain by at least one subclass.

The population II main sequence stars, which lie below the standard main sequence in the color magnitude diagram are often called luminosity class VI objects.

10.3 White dwarf spectra

So far we have only discussed how we can recognize giants and supergiants from their spectra. We have not yet discussed how white dwarfs may be recognized. In Fig. 10.8 we reproduce white dwarf spectra of different spectral types. It is obvious that the spectral lines are extremely broad; in fact, some are so broad that they can hardly be recognized. For most white dwarfs, only hydrogen lines and sometimes the lines of He can be seen. White dwarfs with He lines are called DB stars, while those with hydrogen lines are called DA white dwarfs because, generally, no lines other than hydrogen

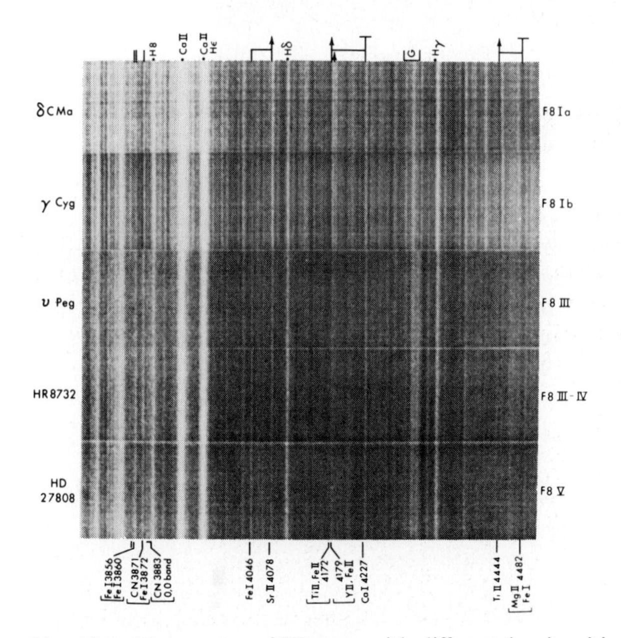

Fig. 10.7. The spectra of F8 stars with different luminosities are shown. The supergiant spectrum (luminosity classes Ia, Ib are at the top), the giant spectrum (luminosity class III) is in the center, the subgiant (luminosity class IV) is below the giant, and the main sequence star spectrum (luminosity class V) is at the bottom. The metallic line strength increases for higher luminosities, especially the strength of the SrII line at 4078 Å. The G band at 4300 Å is broad for giants and main sequence stars. (For Morgan, Abt & Tapscott 1978.)

lines are seen. There are very few white dwarfs for which some of the strongest metallic lines can be recognized, like, for instance, the lines of ionized calcium at wavelengths of 3933 Å and 3968 Å. Some white dwarfs' spectra show no lines at all, and they are called DC spectra or, the stars, DC white dwarfs (C from continuum).

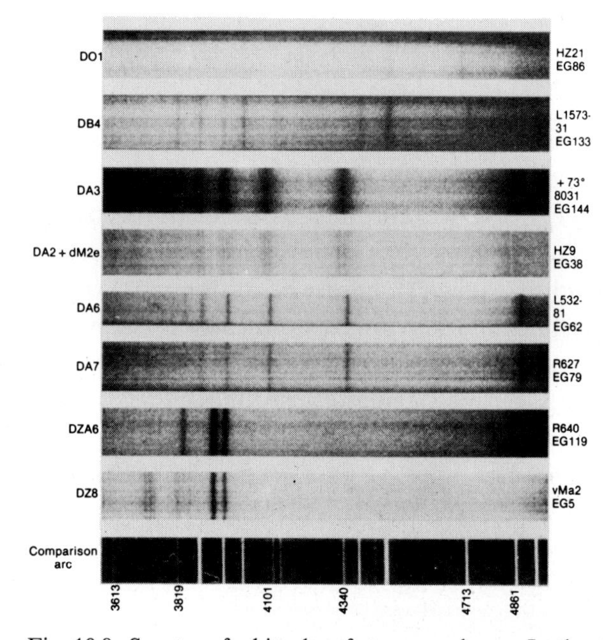

Fig. 10.8. Spectra of white dwarf stars are shown. In the top DO1 and DB4 spectra only helium lines are seen, for instance at 4713 Å and 4471 Å. For the DA star spectra the hydrogen lines at 4101 (H_δ), 4340 (H_γ), and 4861 (H_α) are strong and very broad. In the bottom spectra of R640 and van Maanen 2 the hydrogen lines cannot be recognized. The Ca^+ lines at 3933 and 3968 Å are very broad and strong. (From Greenstein 1986.)

11

Understanding stellar spectra

11.1 The solar spectrum

In Fig. 6.7 we have shown the overall energy distribution of the sun. We have seen that the solar energy distribution has its maximum at a wavelength of about 5000 Å, which, according to Wien's displacement law, would mean that the temperature of the solar surface has to be about 5800 K *if* the sun radiates like a black body (see (4.7)), however, we do not yet know if the sun does, indeed, radiate like a black body. (We saw, however, in Chapter 8, that the effective temperature of the sun is also close to 5800 K). In this section we want to look at the solar line spectrum. In Fig. 11.1 we reproduce a small part of a 'positive' spectrum of the sun, which means we see dark spectral lines at those wavelengths where little or no energy is received. The strongest lines are the lines from Ca^+. This notation means that the lines are due to a calcium atom which has lost one electron and is therefore positively charged by one elementary charge. We call this a calcium ion. The lines are also called Ca II lines, the strongest of which are visible as broad bands at 3933 and 3968 Å. The hydrogen lines seen at 4340, 4861, and 6562 Å are much weaker. Strong lines of magnesium and sodium are also seen as well as many lines of iron, chromium, titanium, silicon, manganese, and other elements.

Fig. 11.1. A small part of the solar line spectrum is shown. Many dark absorption lines, the so-called Fraunhofer lines, are seen. The wavelengths of these lines agree with the wavelengths of spectral lines seen in laboratory spectra of different elements. By comparison with laboratory spectra we can see which elements are present in the sun.

11.2 Line identification

How do we know which atoms or ions cause these spectral lines? We have to study in the laboratory the spectra of light sources which we know contain atoms and ions of a certain element, say, iron. We compare spectra of such a light source with the lines which are seen in stellar spectra or in the solar spectrum. If we see the same lines which we see in the laboratory iron spectrum we know that these lines in the stellar spectrum are caused by iron, at least if we know that no other elements are present. If other elements are also present, we have to study another light source which contains only atoms of that other element in order to identify the lines which do not belong to iron. It has taken a long period of laboratory studies to disentangle all the stellar spectra and to identify which lines are due to which element. We still cannot identify all spectral lines, either in the laboratory or in the solar spectrum. But for the majority of lines we know which element causes them. There are large tables available, for instance the tables by Mrs Moore Sitterly, which list the wavelengths of the lines and give their identification in terms of which atom or ion is responsible for a given line. Sometimes several different elements form a line at nearly the same wavelength. One then has to study in more detail which other lines can be seen. If, for instance, an iron line is a possibility, we have to check whether other strong iron lines are also there; if not, then the identification as an iron line is rather unlikely. In some sense, the identification of spectral lines in a stellar spectrum is like a puzzle: all the pieces have to fit together.

The presence of spectral lines of a given element in the solar or in a stellar spectrum tells us that a particular element is present on the star, but does the line strength tell us how much of that particular element is present? For instance, does the high strength of the Ca^+ lines in the solar spectrum tell us that the sun consists mainly of calcium? Or that the A stars are mainly made of hydrogen? In order to answer this question, we have to discuss in detail the origin of the spectral lines and the factors which determine their strengths. A full discussion of the stellar spectrum analysis must be left until Volume 2. Here we can give only the main results.

11.3 Understanding the spectral sequence

11.3.1 General discussion

In Chapter 10 we saw that most of the stellar spectra can be classified in a two-dimensional scheme, the one dimension given by the spectral sequence and the other by the luminosity classes. We suggested earlier that the spectral sequence is indeed a temperature sequence, while the luminosity classification describes mainly the size of the star (actually the gravitational acceleration in the photosphere). The temperature sequence was suggested because the spectral types are strongly correlated with the colors of the stars. On the other hand, we also saw that different lines and also lines of different elements are seen in different spectral types. Does this suggest that stars with different temperatures have different chemical compositions? A detailed analysis of the stellar spectra shows that this is not the case. Almost all stars in our surroundings have the same chemical composition. How, then, can we understand the changing line strengths of the lines of different elements for stars with different temperatures?

11.3.2 The changing strengths of the hydrogen lines for different spectral types

We discussed in Chpater 10 the fact that the hydrogen lines show maximum strengths for stars of spectral types A0, and decrease their strengths for earlier spectral types (B and O stars) as well as for later spectral types (F, G, K, and M stars). The reason for the decrease of the hydrogen line strengths for the later spectral types is the decreasing temperature for the later spectral types. Only hydrogen atoms in the second quantum level of the hydrogen atom can contribute to the hydrogen line absorption in the visual spectral region. The hydrogen atoms have to be excited into this level and it takes 10 eV to bring the atoms into this excited level. The excitation can occur due to collisions with electrons which need to have high energies for this to happen and there are not many electrons with energies high enough if the temperature is too low. Excitation can also occur due to the absorption of a photon with a wavelength of 1216 Å. If you remember the wavelength dependence of the Planck function you will realise that for low temperatures there are very few photons with such short wavelengths. For decreasing temperatures we therefore find a decreasing number of hydrogen atoms in the excited, second quantum level which can absorb in the hydrogen Balmer lines which we observe in the visual spectral region.

The reason for the decreasing strengths of the Balmer lines at higher temperatures is more difficult to explain briefly. People tend to say that this is due to the ionization of hydrogen which also reduces the number of hydrogen atoms in the second quantum level, because many hydrogen atoms lose their electron altogether. While this is true it is not the real reason for the decreasing line strength. What really counts is the ratio of the number of atoms absorbing in the hydrogen lines to the number of atoms absorbing in the underlying continuum, which for the hot stars are the number of hydrogen atoms in the third quantum level of hydrogen. This number also decreases with increasing temperature in the stellar surface layers, but with increasing temperature their number decreases more slowly than those in the second quantum level, therefore the ratio of those in the second quantum level to those in the third level decreases and this reduces the strength of the Balmer lines.

11.3.3 The strengths of the helium lines

The helium lines which can be observed in the visual spectral region need even higher excitation energies than the hydrogen lines. They can only be excited by collisions with electrons which have at least about 20 eV energy or by photons which have wavelengths as short as about 500 Å. Very high temperatures are required for such high energies of the electrons or to produce a measurable number of photons at such short wavelengths. Temperatures higher than 10 000 K are required to make the helium line visible. For low temperature stars we can never see the helium lines in the absorption spectra of the stars. You may wonder why then the element got the name 'Helium' which means the element only seen in the sun? The sun obviously is a cool star, too cool for helium lines to be seen in the photospheric absorption spectrum. The helium lines were observed in the solar spectrum during solar eclipses when the outer layers of the sun become visible. These are not usually seen because the solar disk spectrum normally seen is so much more luminous than the spectrum which becomes visible during solar eclipses.

The spectrum seen during an eclipse is an emission line spectrum emitted by the high layers above those layers which we normally see. In these high layers, called the solar chromosphere and corona and the transition layer between the chromosphere and corona, the temperature increases outwards and reaches values high enough for the higher energy levels in the helium atom to be excited and then emit light in the helium lines which can be observed. (There is actually also one very weak infrared line of helium which is seen in the normal solar spectrum, but which is also caused by the helium atoms in the high chromospheric layers.) The absence of helium lines in cool stars makes it very difficult to determine helium abundances in cool stars.

For the lines of the helium ions to become visible we need still higher energies first to ionize the helium, which requires about 24 eV as compared to the 13.6 eV required for the ionization of hydrogen, and in addition we need to excite an energy level in the helium ion from which the ions can absorb the visual lines. These lines are therefore seen only for temperatures higher than about 30 000 K.

11.3.4 The strengths of the metallic lines

The lines of the heavier elements are generally expected to be weaker than those of hydrogen because these elements turn out to be very rare in stars. For each 1000 hydrogen atoms we find one atom which is heavier than helium. For high temperatures, when the helium and hydrogen atoms are excited to the energies necessary for the line absorption, the 'metallic' lines are therefore generally much weaker than the helium and hydrogen lines, at least in the visual spectral region; only at lower temperatures, when the hydrogen atoms are not excited to the second quantum level in large enough numbers do the lines from the heavier elements become comparable in strength. For very low temperatures they may even become stronger. For the K stars we see mainly the lines of the neutral atoms because the energies of the free electrons and the energies of the available photons are not high enough to remove one electron from these heavier elements. For somewhat higher temperatures the photons and electrons become more energetic and in those elements from which the electron can be removed easily they can then be stripped from the atoms; they become ionized and we start to see the lines from the ions of iron, titanium, chromium etc. for the G and F stars. The lines of the atoms become weaker because their number decreases.

For A stars, the temperatures become high enough so that now even the ions can again be ionized; they can lose an additional electron and become doubly ionized. The ionization can proceed even further for the B stars and the still hotter O stars. We may then also see lines of Si or Fe which are doubly ionized, the so-called Fe III or Si III lines. Most of the lines of these ions with higher degrees of ionization can only be seen in the ultraviolet, but a few lines are also in the visual spectral region and can be studied to determine the abundance of these elements in the hot stars. In a spectrum analysis of all these different types of stars, the changes of the continuous absorption and the changes in the ionization and excitation have to be taken into account in order to determine the element abundances.

If we do this, we find that nearly all the stars in our neighborhood and actually in our stellar system, our galaxy, have nearly the same chemical composition. We even find that the relative abundances of the heavy elements agree rather well with the relative abundances found originally on earth, especially if we take into account that some of the elements which can form volatile substances have been lost earlier from the Earth's atmosphere.

11.3.5 The molecular bands

In Chapter 10 we discussed the fact that for very red, presumably very cool stars, the M, N, S stars, we also see strong molecular bands which disappear for higher temperature stars.

The difference between atomic lines and molecular bands is due to the large number of energy levels which molecules have. For each excited state of an electron within the molecule, there are an infinite number of rotation and oscillation energy states. The energy differences between the rotational energies are very small. Since spectral lines can originate due to transitions between all these many very close rotational energy states, these lines have only very small wavelength differences. They lie very close in the spectrum, forming a band which can be resolved into separate lines only if very high resolution spectra are taken. On low resolution spectra, the many close lines appear as broad bands.

Molecules form when two atoms or a molecule and an atom collide. They may also be destroyed by photon absorption or in collisions, in the same way as atoms are ionized when they collide with an electron or absorb a photon.

As for atoms, higher temperatures lead to more frequent destruction of molecules and the equilibrium shifts from a large number of molecules to larger numbers of atoms. The binding energy of most molecules like H_2O, TiO, ZrO, CN, etc., is of the order of 4–6 eV. Molecules can, therefore, be destroyed at rather low temperatures. For stars with temperatures higher than about 5000 K, molecular bands become very weak. For the sun, they can only be observed with high resolution, when weak bands can still be detected. In the sun, molecules are only present in the coolest outer layers, and even then they are present only in small numbers. The presence of absence of molecules in the stellar spectra tells us mainly about the temperature of the star, though higher pressures favor the formation of molecules. For most stars the strengths of the molecular bands are quite consistant with the element abundances also observed for the sun.

In Table 11.1 we list the element abundances determined for different stars and compare them with element abundances determined for chondritic meteorites. We see that the relative abundances of the heavy elements determined for the different objects agree rather well. The average abundances determined for the Earth depend on the assumptions made about the interior of the Earth for which we can not make a direct chemical analysis. The best we can tell is that the heavy element abundances also agree with those found in stars if we take into consideration that volatile atoms and molecules have been lost.

Table 11.1 *Logarithmic Element abundances for the most abundant elements determined for stars in our neighbourhood and in the carbonatious chondrites (meteorites).*

Atomic Number	Element	Population I			Population II	
		Sun[2] G2V	α Lyr A0V	Meteorites, carbonaceous chondrites	HD 140 283	HD 19 445
1	Hydrogen (H)	12.0	12.0		12.0	12.0
2	Helium (He)	(11.0)	(11.4)		?	?
8	Oxygen (O)	8.8	9.3		?	?
6	Carbon (C)	8.5	?		6.4	?
7	Nitrogen (N)	8.0	8.8		?	?
10	Neon (Ne)	(7.9)	?		?	?
26	Iron (Fe)	7.6	7.1	7.6	5.2 ± 0.3	5.7 ± 0.2
14	Silicon (Si)	7.5	8.2	7.6	5.1	6.0
12	Magnesium (Mg)	7.4	7.7	7.6	4.9	6.5?
16	Sulfur (S)	7.2	?	7.2	?	?
13	Aluminum (Al)	6.4	5.7	6.5	3.7	4.5
28	Nickel (Ni)	6.3	7.0	6.2	4.2	4.7
20	Calcium (ca)	6.3	6.3	6.4	4.0	4.8
11	Sodium (Na)	6.3	7.3	6.2	3.5	?
24	Chromium (Cr)	5.9	5.6	5.6	3.6	3.8
17	Chlorine (Cl)	5.6	?	4.8	?	?
15	Phosphorus (P)	5.5	?		?	?
25	Manganese (Mn)	5.4	5.3	5.5	2.8	3.7
22	Titanium (Ti)	5.1	4.1		2.8	3.4
27	Cobalt (Co)	5.1	?		2.7	?
19	Potassium (K)	5.0	?	5.1	?	?

1. Normalized to $\log N(H) = 12.00$
2. These abundances are considered to be 'Cosmic Abundances.[1]'

It is found that the vast majority of stars consist mainly of hydrogen with about $10\% \pm 5\%$ of helium mixed in, if we count the number of atoms or ions. Since each helium nucleus is four times as heavy as a proton, the abundance of He by weight, Y, can be computed from

$$Y = \frac{4 \cdot \mathrm{ab(He)}}{4 \cdot \mathrm{ab(He)} + 1\,\mathrm{ab(H)}} = \frac{4 \cdot 0.1}{0.4 + 0.9} = \frac{0.4}{1.3} = 0.31 \pm 0.1$$

The next most abundant elements are C, N, O, and Ne. These four elements together make up about 1/10 of 1% of the total number of particles. Since their atomic weight is, on the average, about 15, we find that their fraction by mass, called Z, is about

$$Z = 0.02 \pm 0.01$$

The next most abundant elements are Si, Mg, Fe, and Al, though these elements are more rare than C, N, O, and Ne by another factor of about 10. We realize that the heavy elements which make up our Earth constitute only a very small fraction of the gas in the cosmos.

Generally the relative abundances of the heavy elements with respect to one another are very similar for all the stars and even for our Earth, though for cool, luminous stars we may find somewhat different relative abundances especially of C, N, and O.

There are, however, a few stars for which the abundances of the heavy elements as compared to hydrogen are very different from those seen in the sun. The overall abundance of the elements heavier than He may be lower by two or even almost three powers of 10. These are the so-called population II stars which we will discuss in the next chapter.

12

Population II stars

There are a few stars in our neighborhood whose spectra show a different chemical composition for their photospheres. These stars were previously known as subdwarfs. The reason for this name was their position in the color magnitude diagram: they appeared below the main sequence, which means they either are too faint for their color or they are too blue for their brightness. A spectrum analysis showed that the latter is the case. It turned out that, for these stars, the relative abundances of the heavy elements with respect to one another are quite similar to the ones observed for the sun, but the overall abundances of the heavy elements with respect to hydrogen and helium are considerably reduced by up to a factor of 500, though most of them have much smaller abundance reductions. In these metal-poor stars, the metallic lines are much weaker than for normal stars of the same temperature. Since spectral lines are generally stronger in the blue, and especially in the ultraviolet, than in the red, the lines take more energy out of the ultraviolet and blue spectral region than out of the red. If the lines are weakened in the metal-poor stars, more energy is restored to the ultraviolet and blue spectral region than in the red and the stars therefore look bluer, especially in the ultraviolet. They show an ultraviolet excess which can, in fact, be used to determine their metal deficiencies. In Fig. 12.1 we show the two-color diagram for the stars in the globular cluster M 92 (object No. 92 in the Messier catalogue of nebulous objects in the sky), and compare it with the two-color diagram of the Hyades star cluster in which the stars have normal or perhaps even slightly higher than normal metal abundances. The cool stars (which have the strong metallic lines) in M 92 have colors which are considerably bluer in the U – B colors than the Hyades stars. The M 92 stars have an 'ultraviolet excess' because they are metal-poor.

For the very metal-poor stars (reduction factors 50–100), the relative abundances of the heavy elements with respect to each other appear to be nearly the same as in the sun. For heavy element reduction factors between 10 and 50, it turns out, however, that different elements have different reduction factors. Apparently, the lighter elements are less reduced than the iron group elements. The differences in the reduction factors, and what this means with respect to the chemical evolution of the stars and the Galaxy, are the subject of many modern studies. A final answer cannot yet be given.

It is interesting to note that these metal-poor so-called subdwarfs also differ from the normal stars in other respects. They all show high velocities relative to the sun. Within the galactic plane, the sun and all the stars in our neighborhood orbit the galactic center. We can also say they rotate around the galactic center the way the planets orbit around the sun. Stars in our neighborhood must orbit with nearly the same speed and direction of velocity as we do, at least if they have nearly circular orbits as the sun has. If stars in our neighborhood have very different velocities, it means that they must have different galactic orbits: they must have very elliptical orbits and/or their orbital planes must be inclined with respect to the orbital plane of the sun, which is identical with the plane of our Galaxy. The high relative velocity of the subdwarfs shows that their orbits are different from the orbits of the 'normal' stars in our neighborhood. They have elliptical orbits which are also inclined with respect to the galactic plane. Indeed, it appears that the overall abundances of the heavy elements in these stars are correlated with the maximum distance z above the galactic plane which these stars can reach (see Fig. 12.2). These very metal-poor stars appear to belong to a stellar population which does not fit into the galactic plane; rather, they belong to a stellar population which originated above the galactic plane. Once they were formed, they fell towards the galactic center. Their angular

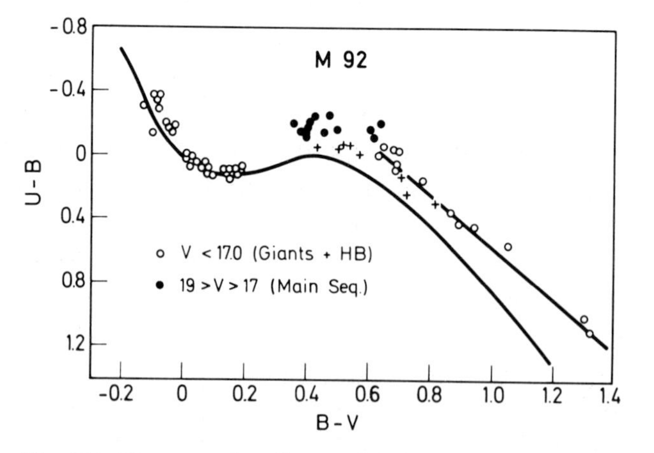

Fig. 12.1. The two-color diagram for the stars in the globular cluster M92 is shown and is compared with the two-color diagram of the Hyades stars. The stars in M92 have more radiation in the U band, which means they have an ultraviolet excess because they are 'metal'-poor.

momentum prevented them from falling into the galactic center, but made them orbit around it in an elliptical orbit, with the orbital plane inclined with respect to the galactic plane (see Fig. 12.2). This population of stars, which does not fit into the galactic plane but, instead, belongs to the space above and around it, the so-called *galactic halo*, are called population II stars in contradistinction to the population I stars which we called the 'normal' stars – those stars with orbits in the galactic plane which also must have been formed within the galactic plane.

In Section 5.3, we discussed the color magnitude diagrams of stars in clusters, and we saw that globular clusters (that is, clusters with a large number of stars and a rather high density of stars), show different color magnitude diagrams from those of the open clusters which have a much lower number of stars. It turns out that the open clusters belong to the population of the galactic plane: their orbits lie in the galactic plane. The globular clusters, on the other hand, belong to the halo population, or the population II. Like the subdwarfs, they also show reduced abundances of the heavy elements. As for the subdwarfs, the most metal-poor globular clusters belong to the largest distances from the galactic plane.

Since it appears that stars form in regions of high densities of interstellar gas and dust, and since we do not see much gas and dust now in the galactic halo, it appears that the population II stars must have formed at a time when there was more gas and dust in the halo than now. This means they must have formed at a time when the galaxy had not yet contracted into a disk. We are, therefore, led to suspect that the population II stars are older than the population I stars, which were formed in the galactic plane. As we just saw, the population II stars are deficient in heavy elements; they must have been formed from gas and dust which was poor in heavy elements. Population I stars, which apparently were formed later, were formed from material which contained a much higher proportion of heavy elements. This suggests that, with time, the interstellar medium from which the stars formed

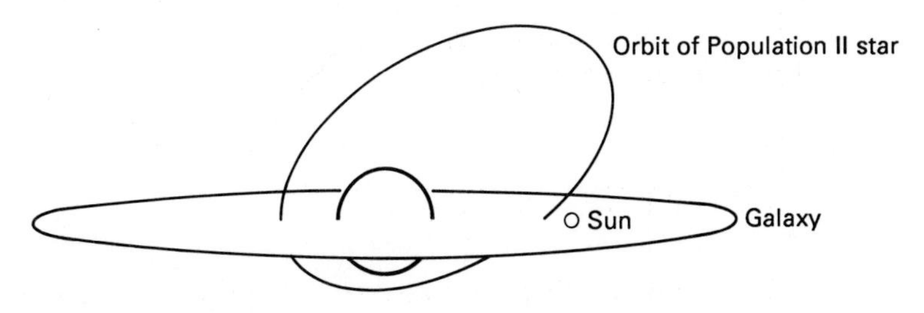

Fig. 12.2. Population II stars have orbits which do not fit into the galactic plane. They generally have elliptical orbits which reach high above the galactic plane into the galactic halo.

was enriched in heavy elements. We expect to find a correlation between the age of the stars and the overall abundances of the heavy elements. We will learn in Volume 3 how we can determine the ages of clusters and stars. Here, we can only state that we have not been able to demonstrate such a correlation, but we also do not find a contradiction. The oldest stars all seem to have about the same age regardless of their chemical composition, at least to within the errors of the age determination. It appears that the enrichment in heavy elements took place in a timespan which is short in comparison with the age of the oldest stars.

13

Stellar rotation

We can measure rotational velocities of stars from the widths of their spectral lines. For a rotating star, one side of the star moves away from us while the other side moves toward us, at least if the line of sight does not coincide with the rotational axis (see Fig. 13.1). The superposition of all the spectra originating from the different parts of the star gives a broadened spectral line (see Fig. 13.2). The width of the broadened line tells us the rotational velocity of the star except for the factor $\sin i$, where i is the angle between the line of sight and the rotational axis.

The effects of rotation on the appearance of the spectral lines is illustrated in Fig. 13.3.

It turns out that the radial velocity, i.e., the component of the rotational velocity in the direction of the line of sight, is constant along stripes across the stellar surface, which are parallel to the direction of the rotational axis projected against the background sky, as we will now show. Suppose the rotational axis is perpendicular to the line of sight. At a given latitude ϕ on the star, the rotational speed is given by (see Fig. 13.4)

$$v_{\mathrm{rot}} = v_{\mathrm{rot}}(\mathrm{equator}) \cos \phi. \tag{13.1}$$

If λ is the geographical longitude on the star counted from the central meridian, then the line of sight component of the rotational velocity, i.e., the radial velocity v_{r} seen for each point on the stellar surface, is given by (see Fig. 13.1(b))

$$v_{\mathrm{rot}} = v_{\mathrm{rot}} \sin \lambda = v_{\mathrm{rot}}(\mathrm{equator}) \times \cos \phi \times \sin \lambda. \tag{13.2}$$

The radial velocity v_{r} is therefore constant for $\cos \phi \sin \lambda = \mathrm{const.}$

If we describe the sphere by x, y, z coordinates in a rectangular coordinate system (see Fig. 13.5), we find, where r is the radius of the sphere,

$$\left.\begin{aligned} z &= r \sin \phi \\ x &= \rho \sin \lambda \\ y &= \rho \cos \lambda \end{aligned}\right\} \tag{13.3}$$

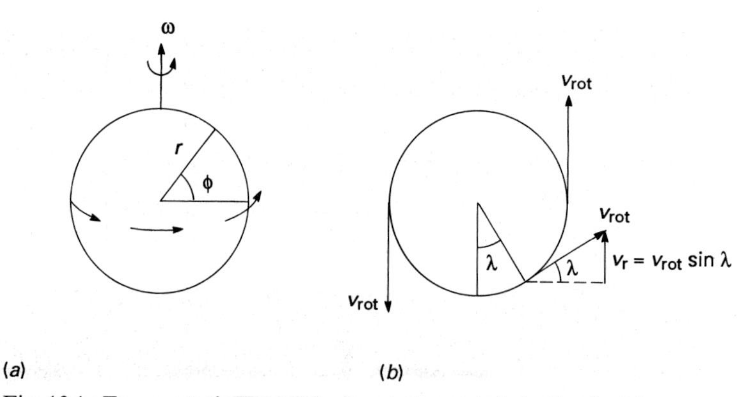

Fig. 13.1. For a rotating star whose rotation axis is inclined with respect to the line of sight, a distant observer will see one side of the star move away from him, while the other side moves towards him.

Fig. (*a*) shows a view of the star with the line of sight in the equatorial plane, Fig. (*b*) a view from an observer whose line of sight is parallel to the rotation axis. An observer whose line of sight is in the equatorial plane sees one side move towards him and the other side move away from him. If the observer looks along the rotation axis no Doppler shift is seen.

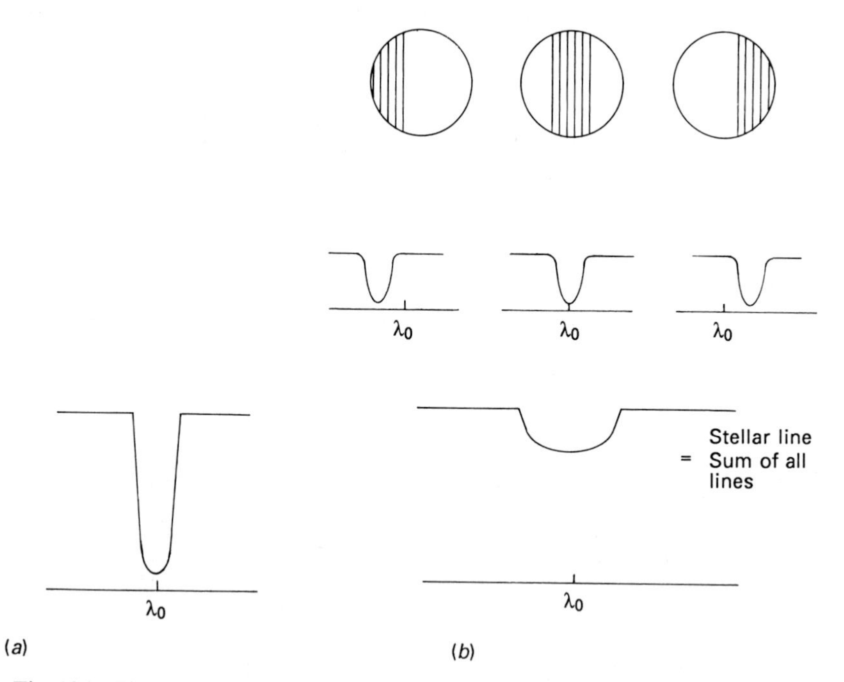

Fig. 13.2. The spectral lines of a non-rotating star (*a*) look generally sharp. For a rotating star, the lines of the opposite sides of the star are shifted in opposite directions, while the center of the star shows the unshifted spectrum (*b*). Since we can not spatially resolve the star, we can only observe the sum of the spectra originating on different parts of the stellar surface. This means we see a broadened line.

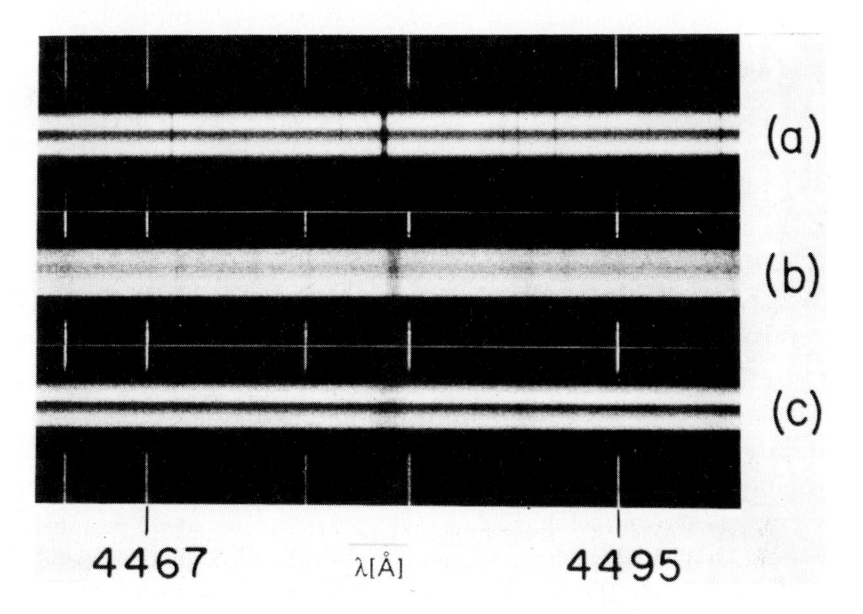

Fig. 13.3. The effects of rotational broadening on the spectral lines are shown for three stars with different rotational velocities. (From Babcock 1960.)

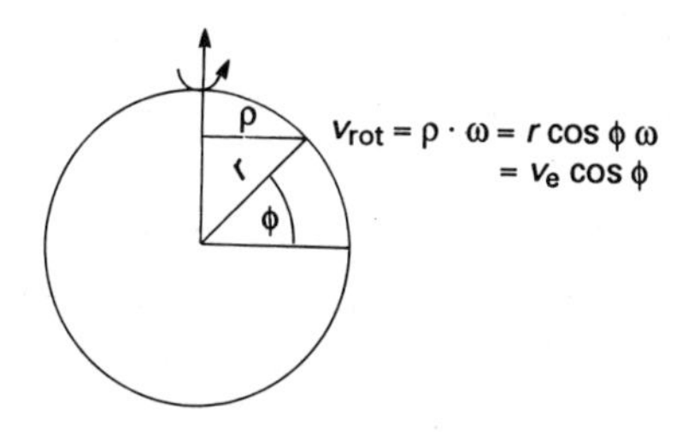

Fig. 13.4. The rotational velocity at the stellar latitude ϕ is given by $v_{rot} = \rho \cdot \omega$ where $\rho = r \cdot \cos \phi$, $v_e = v_{rot}$(equator)

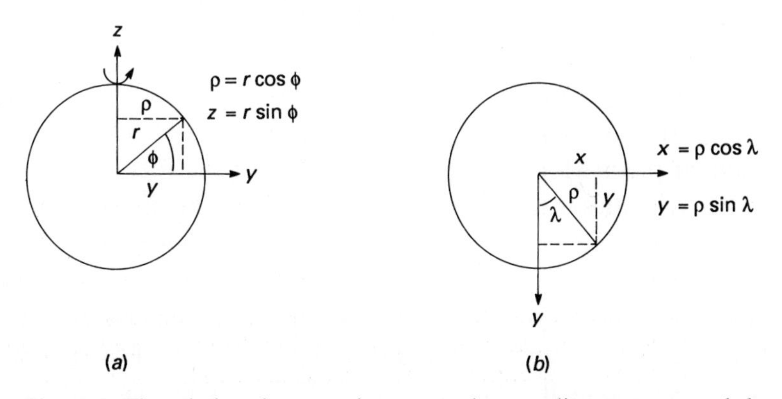

Fig. 13.5. The relations between the rectangular coordinates x, y, z and the spherical polar coordinates r, ϕ, λ are shown. (*a*) View from the equatorial plane. (*b*) View from the pole.

Here, ρ is the radius of the cross section through the sphere at height z. From Fig. 13.4 we see that

$$\rho = r\cos\phi.$$

With this we find

$$x = r\cos\phi\sin\lambda \quad \text{and} \quad y = r\cos\phi\cos\lambda. \tag{13.4}$$

We saw above that v_r = constant for $\cos\phi\sin\lambda$ = constant. From (13.4) we see that this is the case for x = constant on the sphere, which means along cross sections through the sphere at constant distance from the z–y plane. These are stripes on the surface of the star parallel to the projected direction of the rotation axis. In order to calculate the line profile for a rotationally broadened line we have to add up the contributions from these stripes, as shown in Fig. 13.2.

For (13.1)–(13.4) we have assumed that the line of sight is perpendicular to the rotation axis. If this is not the case, all the v_r are multiplied by the factor $\sin i$ where i is the angle between the rotational axis and the line of sight. Generally, we have no way to determine this angle i; we can, therefore, only measure $v_r \sin i$. In the tables giving rotational velocities of stars, we always find this value $v_r \sin i$, which gives us only a lower limit to the actual rotational velocity of a given star.

Assuming a random distribution for the orientation of the rotation axes (probability of i proportional to $\sin i$) we can, however, in a statistical sense correct for the $\sin i$ factor and determine the true distribution of equatorial rotational velocities.

In Fig. 13.6, we show the distribution of average rotational velocities v_{rot}(equator) for field stars as a function of the spectral types. Rather high rotational velocities are found for O, B, and A stars. For late A and early F stars, the rotational velocities drop almost to zero.

In Fig. 13.7, we show the distribution of rotational velocities for stars in different clusters in comparison with the distribution for field stars. While we see a similar picture for all the clusters except for the Pleiades, the drop in rotational velocities is more or less pronounced in the different clusters. In Fig. 13.7, we have actually ordered the clusters according to age. (We will see in Volume 3 how the ages can be determined.) In the old field stars and in the Coma Berenices cluster, the drop in rotational velocities at $B - V = 0.3$ is well pronounced. In the somewhat younger clusters, Praesepe and the Hyades, this drop at $B - V = 0.3$ is less obvious. The rotational velocities remain high also for stars with $B - V > 0.4$. One clearly sees that for stars with $B - V \geqslant 0.3$, a braking mechanism is at work which is even more effective

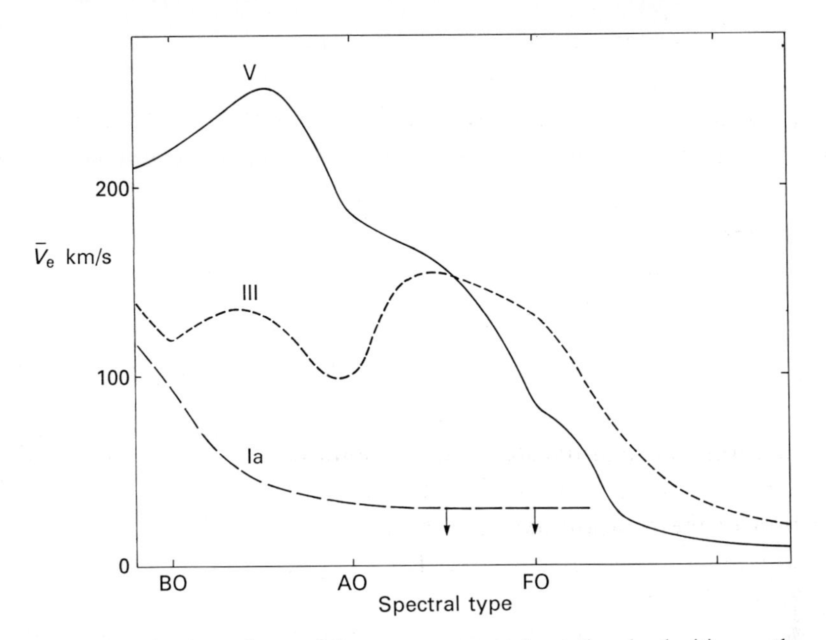

Fig. 13.6. The dependence of the mean equatorial rotational velocities on the spectral types is shown for main sequence stars, for giants, and for supergiants, according to Schmidt-Kaler (1982). V: main sequence stars, III: giants, Ia: supergiants.

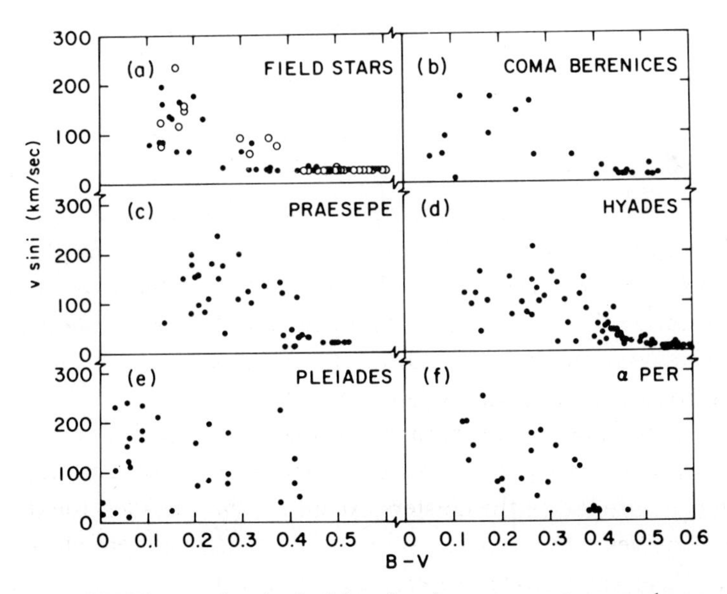

Fig. 13.7. The rotational velocities of main sequence stars are shown as a function of B − V for field stars, i.e., stars which are not in clusters, and for stars in different Galactic clusters. Where necessary, reddening corrections were applied (see Chapter 19). The clusters are ordered according to their ages (see Volume 3). The Coma Berenices, the Hyades and Praesepe clusters are the oldest, the α Perseus and the Pleiades cluster are the youngest. In the Pleiades cluster the stars all still have their large rotational velocities. (From Böhm-Vitense & Canterna 1974.)

for stars with $B-V \geqslant 0.4$. While these subtle differences are not yet understood, astronomers generally believe that this braking mechanism is related to stellar winds, which in turn are related to the presence of stellar coronae which can be observed for the cooler stars. We will come back to this discussion in Volume 2.

14

Stellar magnetic fields

14.1 General discussion

We are all accustomed to the fact that our Earth has a magnetic field whose shape comes rather close to that of a dipole field, with the magnetic axis not being very different from the rotational axis of the Earth. This raises the suspicion that the magnetism is due or at least related to the rotation of the Earth. We may therefore wonder whether stars which have much higher rotational velocities than the Earth might have much stronger magnetic fields than the Earth. It seems very interesting to check. The question arises: how can we measure magnetic fields on stars? We clearly cannot take a magnetometer to the star's surface. All we can get from the star is its light. Fortunately, nature has provided an effect of a magnetic field on the light which can be used to measure magnetic fields of stars. This is the Zeeman effect, named after its discoverer.

14.2 The Zeeman effect

Zeeman discovered that for a laboratory light source which emits an emission line spectrum in a laboratory magnetic field, the spectral lines generally split into several components. If you observe the light in an arbitrary direction with respect to the direction of the magnetic field, you will, in the simplest case, see a line split into three components, the socalled Lorentz triplet. The central component remains at the original wavelength λ_0, which the line had without a magnetic field. The other two components are displaced symmetrically to both sides of the central component. The displacement in frequency, $\Delta\nu$, is given by

$$\Delta\nu = \pm\frac{e}{m}\times\frac{H}{4\pi c} = 4.66\times10^{-5}\times H, \quad \text{with} \quad \nu = \frac{1}{\lambda} \tag{14.1}$$

or, expressed in $\Delta\lambda$, we find with

$$\frac{\Delta\lambda}{\Delta\nu} = \frac{\mathrm{d}}{\mathrm{d}\nu}\left(\frac{1}{\nu}\right) = -\frac{1}{\nu^2} = -\lambda^2$$

$$\Delta\lambda = \pm\lambda^2 \times 4.7 \times 10^{-5} \times H \tag{14.1(a)}$$

where H is the magnetic field strength to be measured in Gauss, e is the electronic charge to be measured in electromagnetic units, and m is the electronic mass.

The relative strengths of the three components change if we observe in different directions. If we observe in a direction parallel to the magnetic field lines, we see what is called the *longitudinal Zeeman* effect. In this case, we see only the two displaced components: the intensity of the central component has decreased to zero (see Fig. 14.1). The light in the two displaced components turns out to be *circularly polarized* in opposite directions. The direction of the circular polarization for each component depends on the direction of the magnetic field. If, in Fig. 14.1, the magnetic field around the light source were to point in the opposite direction, the directions of the circular polarization for the two lines would have to be inverted.

If we were to observe the light of the light source in the direction perpendicular to the direction of the magnetic field, we would observe the so-called *transverse* Zeeman effect, which means we would see all three components of the Zeeman triplet, the central component being twice as strong as each of the displaced components. Now all three components turn

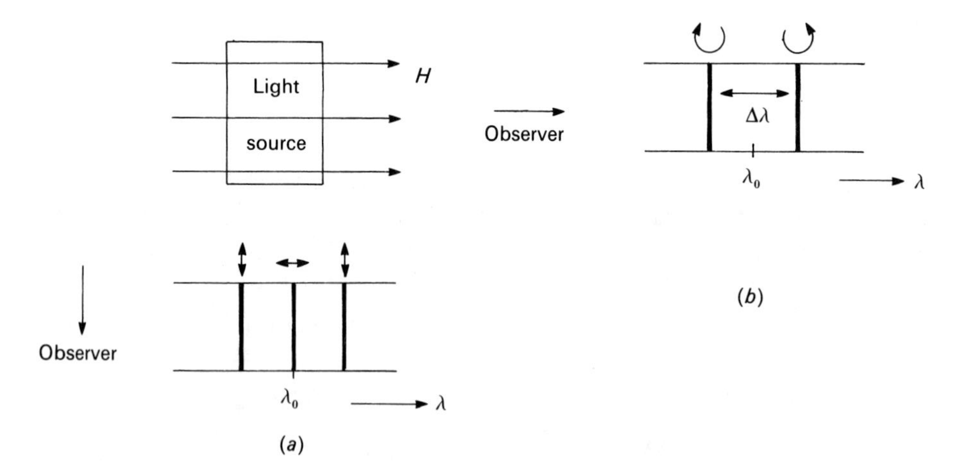

Fig. 14.1. The splitting of a spectral line due to the Zeeman effect is shown for the case where we observe the light source in the direction parallel to the magnetic field, i.e., we observe the longitudinal Zeeman effect (*a*), and for the case when we observe perpendicular to the magnetic field, i.e., when we observe the transversal Zeeman effect (*b*). The directions of circular and linear polarizations of the emitted light are indicated by the arrows.

out to be *linearly* polarized. The two shifted components are polarized in the direction perpendicular to the magnetic field, while the central component is polarized parallel to the magnetic field (see Fig. 14.1, the lower left plot).

Since the splitting of the lines increases with increasing magnetic field, a measurement of the line splitting gives a measurement of the magnetic field strength. Hale, in 1908, was the first to make use of this Zeeman effect to measure the magnetic fields in sunspots. In Fig. 14.2 we show a picture of a solar sunspot, taken in the light of the hydrogen Hα line. This picture reminds us very much of the magnetic lines of force outlined by iron filings on a piece of paper on top of a small magnet in the laboratory (see Fig. 14.3). This made Hale suspicious, and he actually measured magnetic fields of several thousand Gauss in the sunspots.

From (14.1(a)) we can calculate how large the splitting of a spectral line at about $\lambda_0 = 4300$ Å will be. For a field of 1000 Gauss, we find $\Delta\lambda/\lambda = 2 \times 10^{-6}$. Due to rotation, the spectral line of a star has a width of about $\Delta\lambda_D = \lambda v_r/c$, where c is the velocity of light. For $v_r \sin i = 100\,\mathrm{km\,s^{-1}}$, we find a width of $\Delta\lambda = 1.4$ Å. This means that the widths of the spectral lines due to rotation are much larger than the splitting of the lines due to the Zeeman effect. We will not be able to measure the splitting of the line. Only in cases of a small $v_r \sin i$ and in the case of a very strong magnetic field, can we hope to see the splitting of the lines directly. In Fig. 14.4 we show the spectrum of HD 215441 (star No. 215441 in the Henry Draper catalogue of stars), as taken by Babcock 1960. In this spectrum the three Zeeman components of the Cr line

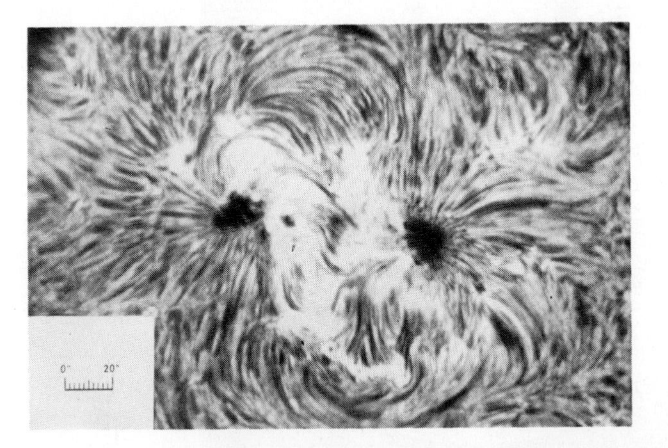

Fig. 14.2. High spatial resolution pictures of sunspots are seen, taken in the light of the H_α line of hydrogen. We have the impression that we actually see the outline of magnetic field lines of force emerging from the sunspots. (From Bray & Loughhead 1974.)

at 4254 Å can actually be seen. The amount of splitting indicates a magnetic field of 34 000 Gauss for this star.

For stars with magnetic fields of the order of 1000 Gauss, as are observed in sunspots, and for rotational velocities of the order of 50 km s^{-1}, the

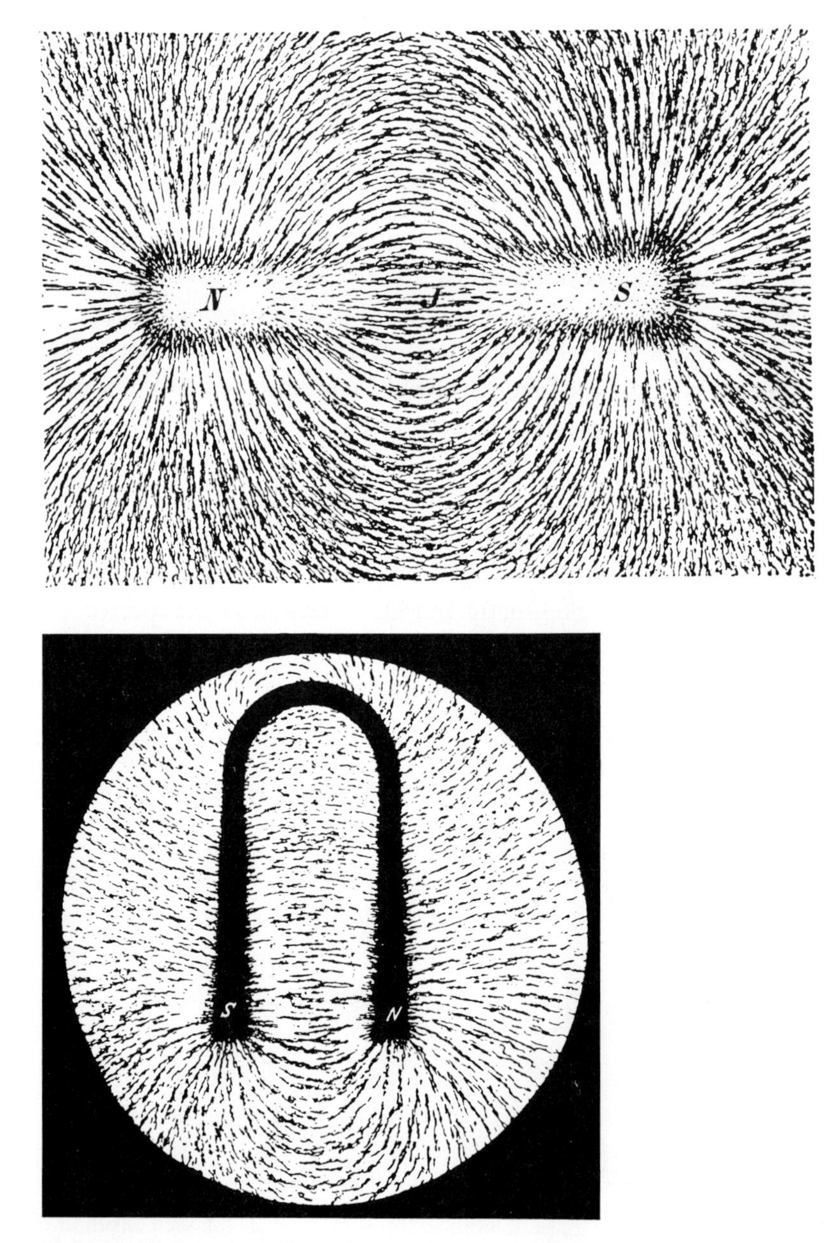

Fig. 14.3. The magnetic lines of force of a laboratory magnet as outlined by iron filings on a piece of paper on top of the magnet (top picture), and on the side of a magnet (bottom picture).

different components cannot be resolved. We observe only slightly broadened lines (see Fig. 14.5). We clearly have the best chance to separate the components if we observe the longitudinal Zeeman effect only, which eliminates the central component. However, we cannot arrange our line of sight. We therefore have to build an instrument which manages to separate the components with circular polarization. Babcock, at Mount Wilson observatory, was the first to design such an instrument for the observation of

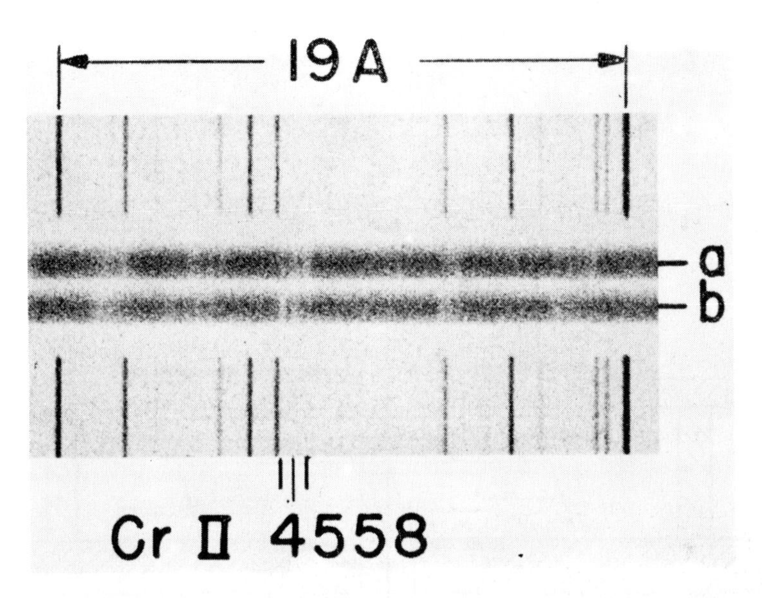

Fig. 14.4. We reproduce a small section of the spectrum of the star No. 215441 in the Henry Draper catalogue. This star has a magnetic field of 34 000 Gauss and a rotational velocity of only 5 km s^{-1}. The lines are narrow enough and the magnetic field strong enough so the Zeeman triplet of the Cr line at 4558 Å can be seen clearly. The laboratory (emission line) spectra seen above and below the stellar spectrum are needed for the line identification. (From Babcock 1960.)

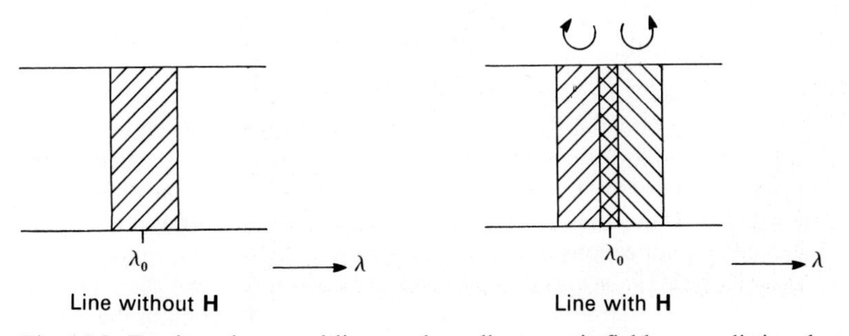

Fig. 14.5. For broad spectral lines and small magnetic fields, no splitting, but only a broadening of the line can be seen due to a magnetic field. The light on the two sides of the line is circularly polarized in opposite directions.

stellar magnetic fields. His instrument is shown in Fig. 14.6. The essential parts are the calcite crystal and the $\lambda/4$ mica. The calcite crystal has the property of having different diffraction angles for linearly polarized light in perpendicular directions. The $\lambda/4$ mica makes linearly polarized light out of circularly polarized light. For opposite directions of the circularly polarized light the directions of the linearly polarized light obtained are perpendicular to each other. A generally unpolarized beam of light from the star, which we can regard as consisting of two beams of circularily polarized light, being

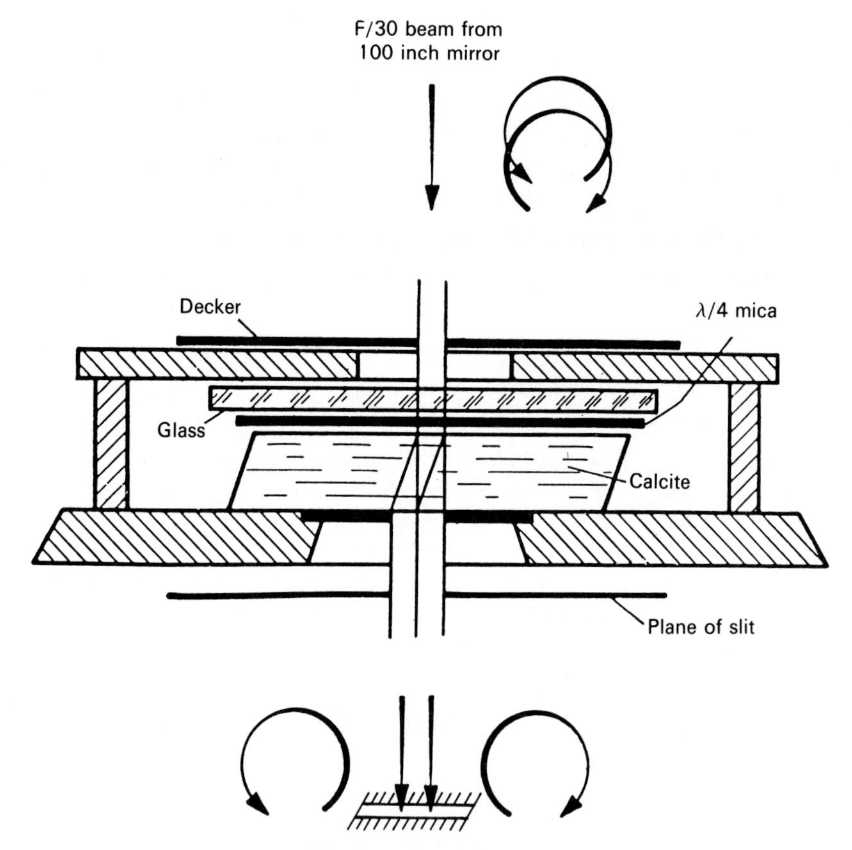

Fig. 14.6. The essential parts of Babcock's magnetograph are shown. The $\lambda/4$ mica plate transforms circularly polarized light into linearly polarized light: opposite directions of circular polarization lead to linear polarization in perpendicular directions. The two coinciding light beams from the star with perpendicular directions of linear polarization suffer different diffraction in the calcite crystal. The beams appear side by side when they exit the crystal. They then enter the spectrograph slit separated along the direction of the slit and form two spectra side by side on the photographic plate, where the spectra of opposite directions of polarization are then seen side by side. (From Babcock 1960.)

polarized in opposite directions, enters the $\lambda/4$ mica plate. On the other side of the mica plate the beam now consists of two linearily polarized beams which have perpendicular directions of polarization. They now pass through the calcite crystal in which one of the beams passes straight through, while the other beam is diffracted. The beams with different polarization therefore exit the calcite crystal at different places. The calcite crystal thickness is such that the two beams exit side by side. The beams then enter the spectrograph slit. In the spectrograph they are each dispersed into a spectrum. The two spectra from the two beams now appear side by side. Since the one beam contained light from the one direction of polarization a given spectral line is shifted, say, to longer wavelengths if the star has a magnetic field. In the other spectrum the light with opposite circular polarization is seen. In this spectrum we see a line which is then shifted to shorter wave-lengths if the star has a magnetic field. In those two side by side spectra the line therefore appears at slightly different wavelengths, if the star has a magnetic field. The shift corresponds to the separation of the two shifted lines from the Lorentz triplet.

For different lines the amount of splitting in a given magnetic field is different, depending on the structure of the energy levels to which the line transition belongs. For some lines the splitting is larger than is given by (14.1), and for some it is smaller. Some lines do not split at all. The sensitivity of the different lines to the Zeeman effect can be measured in the laboratory, and we can check whether the measured splittings for the different lines show the same ratios as in the laboratory. In this way we can make sure that we are actually seeing the effect of a magnetic field on the star.

The line shifts for the two spectra are measurable even if the lines are broader than the shift, as can be seen from Fig. 14.7 and also in Fig. 14.8 where we show such pairs of spectra as observed by Babcock. Even with this

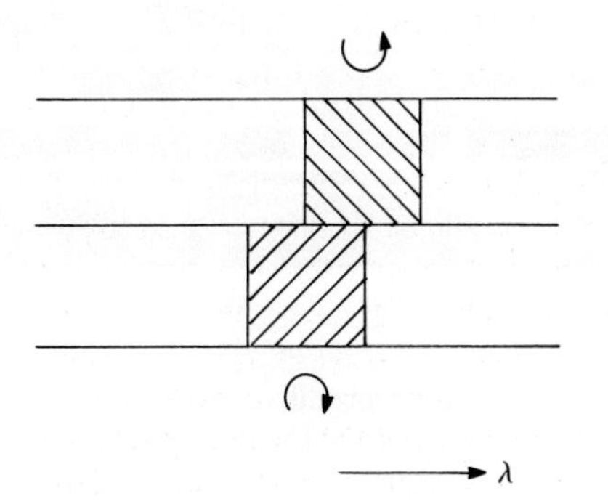

Fig. 14.7. Sets of two spectra of opposite circular polarization are obtained with a magnetograph.

clever instrument, we are limited to fairly strong magnetic fields for our measurements. Fields of less than 100 Gauss cannot be measured even today, though there are still more sophisticated instruments now in use. Modern instruments make use of the polarization created by the magnetic field line shifts in the hydrogen lines. The principle is the same as described here. The shifts of the lines in the polarized beams causes a slight polarization in the wings of the hydrogen lines which has opposite signs in the long wavelength ans short wavelength wings. A difference in polarization for the two wings indicates a stellar magnetic field.

As we mentioned previously, the splitting of different metallic lines due to the Zeeman effect is different. For each line, the splitting can be measured in the laboratory for different field strengths, and the ratio of the splitting for different lines for a given field strength can be determined. As a further check on our interpretation of the splitting as the result of the Zeeman effect, we can determine in the stellar spectra the ratio of the splitting for the different lines seen in the stellar spectra and check whether they agree with the laboratory values. Such checks were done by Babcock and he found good agreement.

It turned out that for most stars for which magnetic fields could be measured (only a small fraction of all stars), these fields vary with time in a periodic way. In Fig. 14.9 we have plotted the fields measured by Babcock for stars with measurable magnetic fields as a function of their spectral types. For the stars with variable magnetic fields, the maximum field strength was plotted regardless of whether we see N or S polarity. From Babcock's measurements it appeared that the maximum field strengths occur for early A stars. No measurable magnetic fields were found for cool stars. We know that the sun has a weak magnetic field which is also periodically variable. We can measure this field of a few Gauss on the sun only because we can measure single spots on the solar surface where the lines are not rotationally

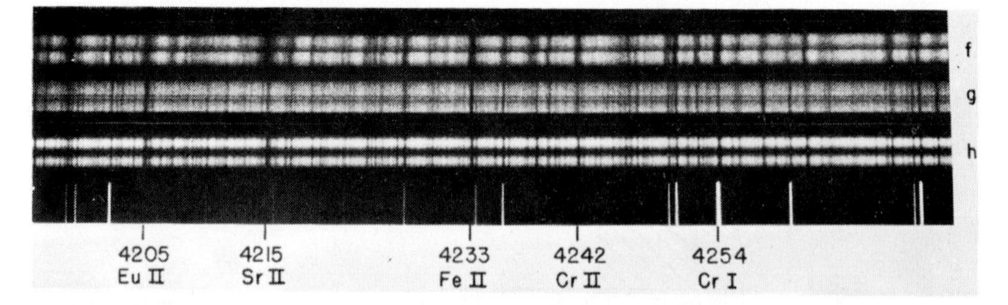

Fig. 14.8. Spectra of three magnetic stars are reproduced, taken with the Babcock magnetograph, showing the spectra of light with opposite directions of circular polarization side by side. The shift of the spectral lines due to the Zeeman effect can be seen especially well for the star 53 Camelopedalis (f). (From Babcock 1960.)

broadened. In addition, we can use very high resolution spectra of the solar light because the sun is so bright. If another star like the sun had a magnetic field comparable in strength to the solar field, we would never be able to measure it. Modern measurements in the infrared do however reveal magnetic fields of the order of about 100 Gauss for some cool stars. For hot stars, which means for O and early B stars, strong magnetic fields were not found by Babcock, because the metallic lines are so weak and there are hardly any stars which have $v_r \sin i$ small enough to permit the measurement of the Zeeman splitting for the metallic lines. The B stars do have, however, strong hydrogen lines. The polarization measurements in the hydrogen line wings have revealed strong magnetic fields for some peculiar B stars, the so-called weak and the so-called strong helium line stars, whose helium line strengths do not match the strengths expected for their colors. The weak helium line stars are mainly late B type stars, while the strong helium line stars are early B stars. It appears that the peculiar late B type stars have in the average magnetic fields comparable to the peculiar A stars (Borra *et al.* 1983), while for the peculiar early B type stars even stronger fields are found (see Bohlender *et al.* 1987). We do not yet know whether there might be peculiar O stars which have even stronger magnetic fields.

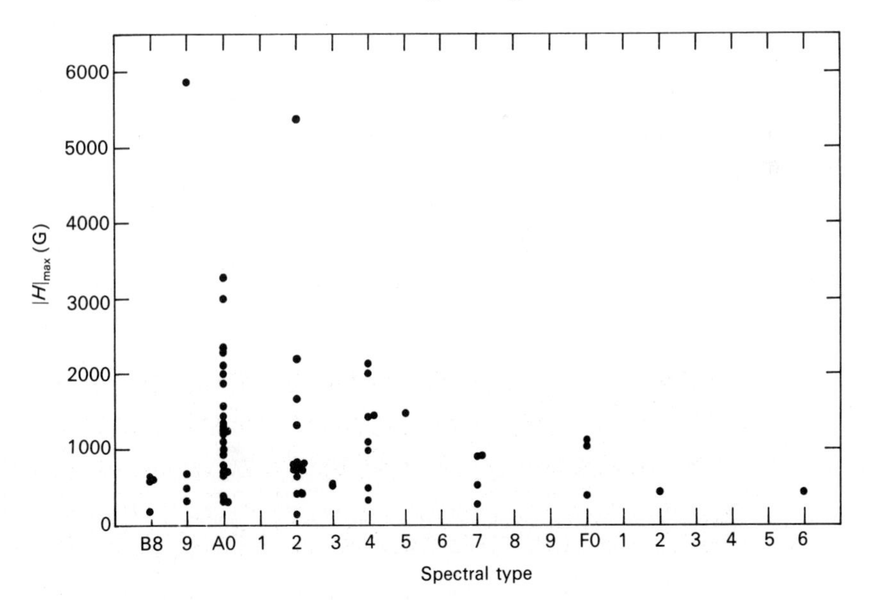

Fig. 14.9. The maximum measured field strengths for stars with magnetic fields (only a small fraction of stars of any spectral types have magnetic fields) are plotted as a function of their spectral types according to the data given by Babcock (1960). The largest fields were measured for early A or late B type stars. Even larger fields are now found for peculiar early B stars with unusually strong helium lines.

15

Stars with peculiar spectra

15.1 General discussion

In the previous sections we have discussed stars which are generally considered to be normal stars, which means their spectra fit into the two-dimensional classification scheme according to spectral type and luminosity. True, the weak-lined stars, or population II stars, do not fit into that scheme, but generally their peculiarity can be understood by the change of just one parameter, the ratio of the metal abundance to the hydrogen abundance, though recently it has been found that this may not always be the case. More than one parameter may actually be necessary to describe the abundances of the heavy elements. The population II stars are still generally considered to be 'normal' stars because we believe that all their peculiarities can apparently be traced back to different chemical abundances. For the stars we are going to discuss in this chapter, this does not seem to be the case. There are, of course, a large number of different kinds of peculiar stars, but we are not able to discuss all of them in the framework of this introduction to stellar astrophysics. We shall only discuss the most frequent kinds of peculiar stars and those which are of special interest in the framework of understanding stellar structure and evolution.

15.2 Peculiar A stars, or magnetic stars

15.2.1 The observations

In the previous section we saw that some stars with very strong magnetic fields are found among the early A stars. These stars were previously known as peculiar A stars or Ap stars because of their peculiar spectra. They show unusually strong lines of Y, Si, Sr, Cr, and Eu, as well as other rare earths. These peculiarities are clearly recognizable on low resolution spectra which are used for classification. In Fig. 15.1 we compare spectra of peculiar A stars with those of normal A stars. On higher resolution spectra it becomes obvious that the lines of nearly all the rare earth elements are stronger than in normal spectra. Spectrum analysis indicates abundances which are increased by factors of up to 1000 for the rare earth elements. Astronomers found it hard to believe that the rare earth elements, especially, should be enhanced by such large factors in these stars. They

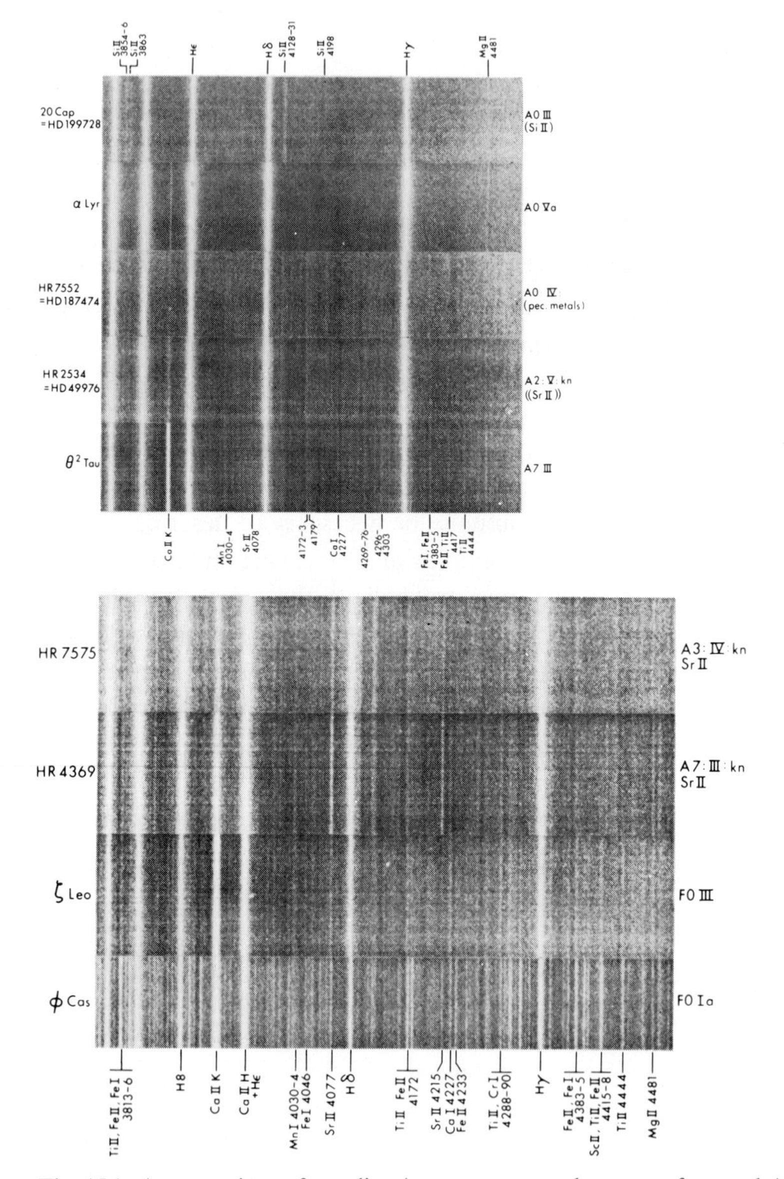

Fig. 15.1. A comparison of peculiar A star spectra and spectra of normal A stars is shown. The peculiar A stars show much stronger lines of silicon ($\lambda = 4128$–31 Å), strontium ($\lambda = 4077$ and 4215 Å), and the rare earths. (From Morgan *et al.* 1978.)

looked for all possible explanations by means of peculiar structures of the photospheres which might explain the peculiar line intensities, but no explanation was found.

A peculiar structure of the photospheres was actually suggested by the fact that the line intensities in these stellar spectra are periodically variable in time. The chemical abundances can hardly be expected to be variable in time. If, however, different parts of the stellar surface have different photospheric structures, perhaps due to the influence of magnetic fields, then because of stellar rotation, we may see different parts of the surface at different times. This may explain the variations; that is, if different parts of the stellar surface have different chemical compositions, we expect, of course, similar variations in their spectra. The period of line intensity variations must then be the period of rotation. There was only one difficulty, namely, that some stars showed different periods of variation for different lines. Some lines showed periods which were shorter by a factor of two. In Fig. 15.2, we show the line intensity variations for the lines in the peculiar A star 56 Arietis. We see that the period for the variation of the line strengths for the helium lines is half as long as the period for the Fe and Si lines.

In 1950, when Babcock started to measure stellar fields, he found that all the A stars with magnetic fields were peculiar A stars, and all the Ap stars

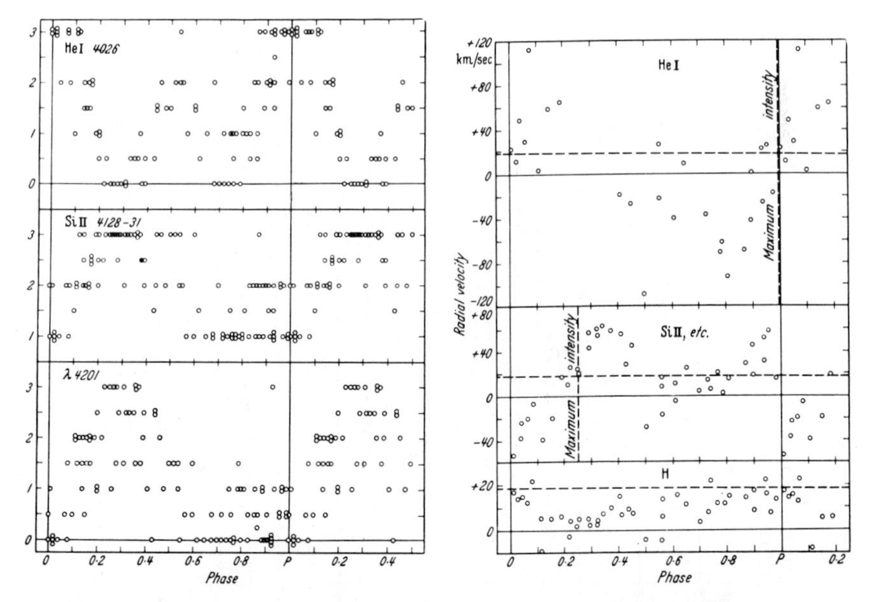

Fig. 15.2. For the peculiar A star 56 Arietis the line strengths of different lines are shown as a function of phase. For some lines the period appears to be half as long as for other lines, for instance, for those of helium. (From Deutsch 1958.)

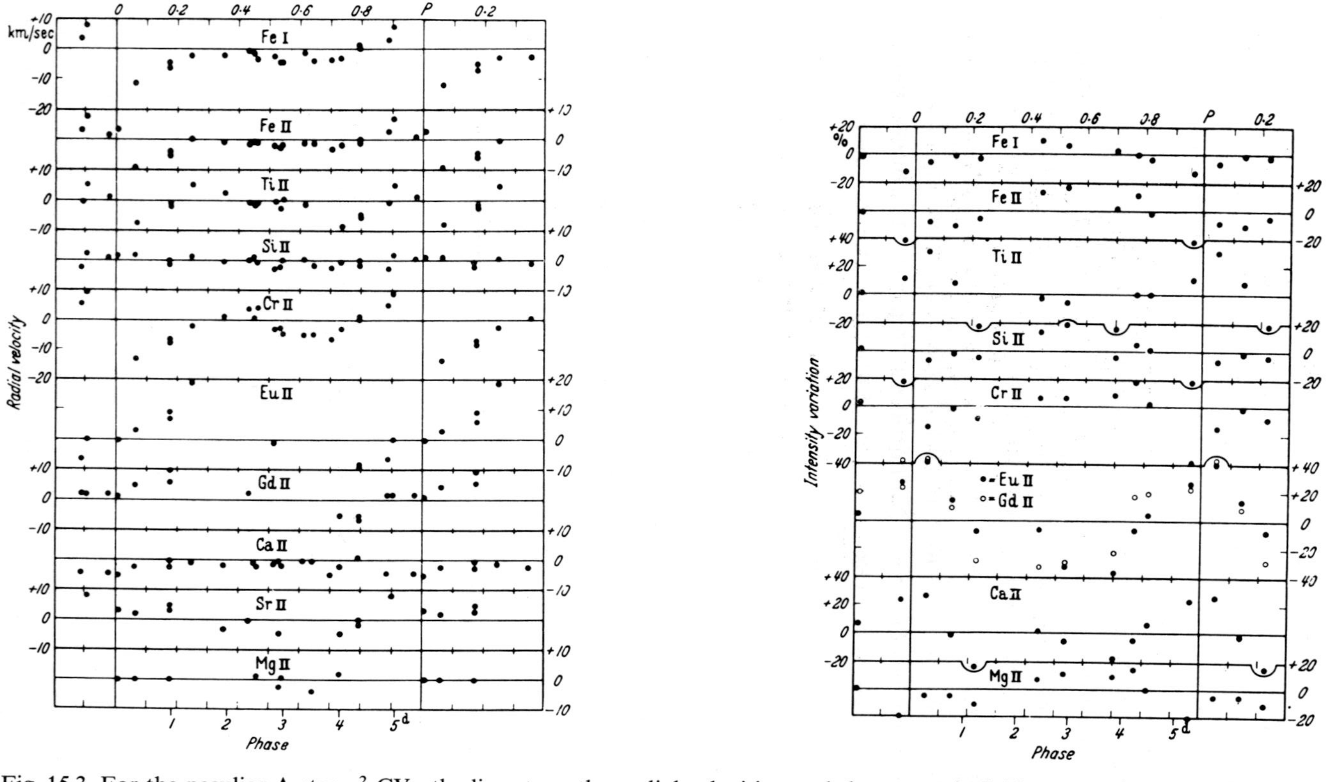

Fig. 15.3. For the peculiar A star α^2 CVn the line strengths, radial velocities, and the magnetic fields are shown as a function of phase. If a given line strength increases, its radial velocity is negative. For decreasing line strength the radial velocity is positive. (From Ledoux & Walraven 1958.)

whose $v_r \sin i$ was small enough, and whose lines, therefore, were narrow enough for the Zeeman effect to be measurable, had strong magnetic fields. We can, therefore, conclude that all Ap stars have strong magnetic fields. As we said earlier, the magnetic fields for many of these stars are also periodically variable, and it was found that the periods agree with the spectroscopic periods, i.e., with the longer periods of line strength variation. Radial velocities measured from different lines in the stellar spectra also turned out to be periodically variable, but different kinds of variation are observed for different lines. For phases of increasing line strength the radial velocity is generally negative, which means the material is moving towards us. For decreasing line strength the material is generally moving away from us. In Fig. 15.3 we show the observed variations of the line strengths, the radial velocities, and the magnetic field for the star α^2CVn in the correct phase relations. The observed line intensities and radial velocities are very nearly what we would expect for an oblique rotator (see next section) if lines of different elements have different line strengths on different parts of the stellar surface.

15.2.2 *The oblique rotator hypothesis*

If we assume that some lines are especially strong on one part of the stellar surface, then the rotational velocity of that spot is directed toward us when rotation brings it around the stellar limb into the visible hemisphere (see Fig. 15.4). When the spot disappears on the opposite limb, the radial

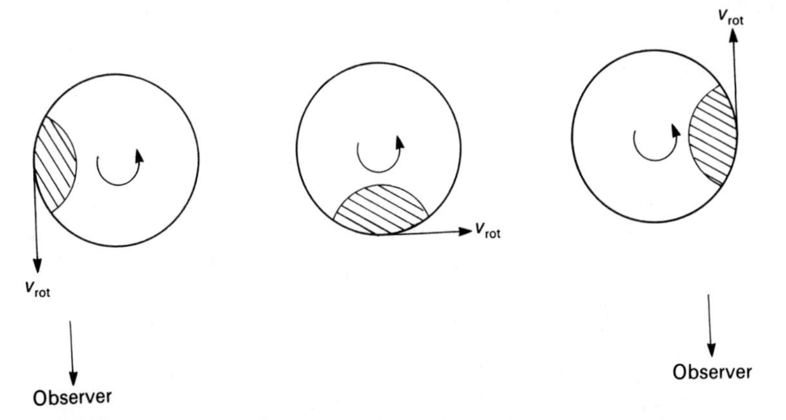

Fig. 15.4. If rotation brings a spot with especially large line strength into the visible hemisphere, i.e., when the lines become stronger, the radial velocity of that line is directed toward us, which means it is negative. If the spot disappears on the other limb and the line becomes weaker, the radial velocity is directed away from us, which means positive.

velocity is directed away from us, which means it is positive. With this oblique rotator hypothesis, we can also understand why some line intensities have half the period of others. These lines originate in two spots on the star and, therefore, have two maxima. We call this model the *oblique* rotator model because rotation leads to changes in the magnetic field strength only if the magnetic axis is inclined with respect to the rotation axis (see Fig. 15.5). If we assume that the magnetic field of the star is similar to a dipole field, then we can find a reversal of the polarity of the magnetic field, as is measured for several of the Ap stars: we sometimes look at one pole and sometimes at the other. If the magnetic axis and the rotation axis were parallel, we would always have the same aspect of the magnetic field, at least if the magnetic field was symmetric around the magnetic axis.

Another question is whether the magnetic field is, indeed, similar to a dipole field. Stibbs calculated that, for a dipole field and the oblique rotator model, we should always observe sinusoidal variations of the magnetic field. Babcock's measurements did not show sinusoidal field variations, but new measurements made with the hydrogen line polarization method suggest that the variations may, indeed, be sinusoidal and that the measurements of metallic line Zeeman splitting may lead to somewhat erroneous field variations because of the changing line strengths on different parts of the stellar surface which have different magnetic field strengths. The shape of the magnetic field is, at present, not very well known. One fact, however, seems to be well established: the polar strengths of the field are not equal at both poles for many of the stars. It appears that the field consists of at least two parts: one nearly dipole field and one nearly quadrupole field, as illustrated in Fig. 15.6. The sum of the two shows different field strengths at both poles because at the one pole, the dipole and quadrupole fields add up, but at the

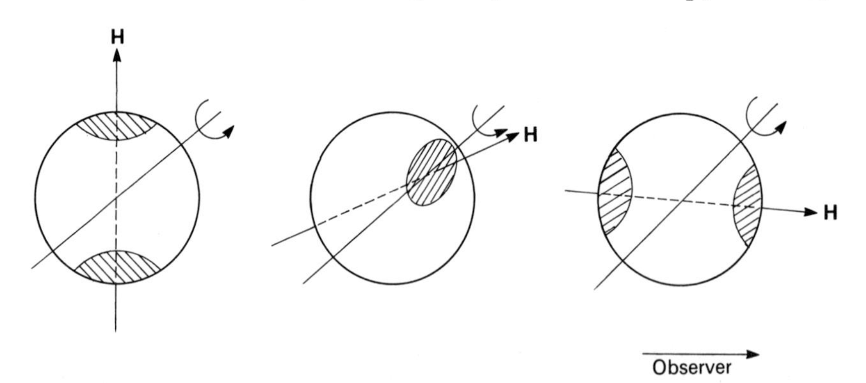

Fig. 15.5. Only if the rotation axis and the magnetic axis are inclined with respect to each other does the rotation lead to changing aspects of the magnetic field, and thereby to changing values of the measured magnetic field strength.

other pole they have opposite sign and the difference of the field strengths is observed. With such a field, symmetric around the magnetic axis and an inclined rotation axis, all the observed variations can be understood quite well. We conclude that the lines which show half the period lengths of the rest of the lines originate at both polar regions, while the other lines must be less strong at the poles and stronger in the equatorial belt, where 'equatorial' means 'with respect to the magnetic axis.'

The oblique rotator hypothesis can be checked in yet another way. Stars with short periods of rotation must generally have broad lines unless we happen to see them nearly pole on (here we mean the rotational poles). Those with long rotation periods must have narrow lines. If, indeed, the periods for the line strength variations are the rotation periods, then stars with short period variations must generally have broad lines, while those with longer periods should have narrow lines. This test was first done by Deutsch (1952) at Mt. Wilson, and the correct relation was found. While this does not necessarily prove that the oblique rotator hypothesis is correct, it does prove that this hypothesis is a viable possibility to explain all the observed variations.

How else can we understand the periodic variations of the magnetic field with periods of the order of days to weeks? Geologists are talking about reversals of the Earth's magnetic field, though not within weeks, but nevertheless it seems to have happened over very long time intervals.

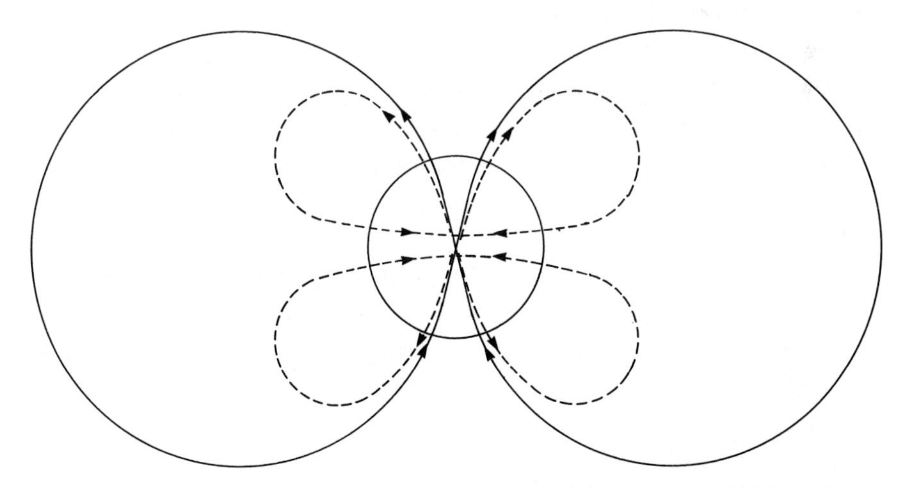

Fig. 15.6. The magnetic field strengths measured by Babcock at the two poles of a magnetic star are not equal. We then have to assume that the actual field is the sum of a dipole and a weaker quadrupole whose axis is aligned with that of a dipole. At the one pole the field strengths add up, while at the other they are of opposite polarity and, therefore, only the difference in field strength is observed. The two poles then are observed to have different field strengths.

We also know of reversals of the general solar magnetic field which happen about every 11 years, not within a week either, but slowly and irregularly over several years. Could that also happen with the magnetic stars?

In order to answer this question, we have to remember that the general magnetic field of the Earth, and also of the sun, is of the order of one Gauss. We also know that matter and magnetic fields can only move together, but which controls the motion, the matter or the magnetic field? We can judge that when we compare the magnetic energy of the field, which is a measure of the resistance of the field to any changes, with the kinetic energy of the gas. It is easiest to compare the magnetic pressure P_M with the gas pressure P_g.

In a homogeneous magnetic field, the magnetic pressure is

$$P_M = \frac{H^2}{8\pi} \tag{15.1}$$

where H is the magnetic field strength.

For a field H of one Gauss, the magnetic pressure, and therefore the magnetic forces, are very small. In the solar atmosphere, the gas pressure P_g is on the order of 10^5 dyn cm^{-2}. The kinetic energy of the matter is much larger than the magnetic energy. The material controls the motion and the changes in the magnetic field, at least at the surface. For a field of 1000 Gauss, the situation changes. We now find $P_M = 10^6/8\pi \sim 10^5$. Now, magnetic energy and kinetic energy are comparable. In sunspots, the magnetic field determines the structure. In A stars, the gas pressure in the atmosphere is only 10^3–10^4 dyn cm^{-2} (the absorption coefficient is so large that we only see the top layers where the gas pressure is low). For a field of several 1000 Gauss, the magnetic energy is much larger than the kinetic energy. The motion of the material is not able to control the structure of the magnetic field at least not at the stellar surface.

Nevertheless, let us consider one possibility of varying magnetic field strength which has been discussed in the literature. In order for the magnetic field to increase in strength, the material has to squeeze the field lines together (see Fig. 15.7). On a magnetic star, one can imagine that the field strength increases when the equatorial material moves toward the poles, which would give an increase in the field strength near the poles and a decrease near the equator (see Fig. 15.8). However, we see from Fig. 15.8 that, while we can get a change in the magnetic field in this way, we can never get a reversal of polarity, as is observed. While this was first discussed as one possibility to explain the magnetic field variations, most astronomers are now convinced that this mechanism does not work.

The other possibility is, then, an oblique rotator (see Fig. 15.5). If the axis of rotation is inclined close to 90° and the star has a dipole field, the observer sees a sinusoidal variation of the magnetic field with changing polarity. The period of variation is the period of rotation. If some spectral lines are stronger at the magnetic poles, we expect two maxima of line strengths during one magnetic period.

But why are some lines stronger at the poles and some stronger at the equator? In the beginning, it was discussed whether nuclear reactions could take place in the polar regions, but the observed abundance ratios do not agree with those expected from nuclear reactions. There are also difficulties in accelerating particles to energies high enough for nuclear reactions: In order to obtain high velocities the densities have to be low enough so that collisions do not occur, but at such low densities, nuclear reactions do not occur either. There is also the peculiar observation that the enhancement of line strengths depends on the effective temperature of the stars. For the hotter stars, we see strong Si lines; the cooler stars have strong Eu, Sr, and Cr lines. It would be peculiar if the nuclear reactions were dependent on the effective temperature of the star rather than on the magnetic field strength of the star. *Diffusion* is considered to be the only possible explanation for the

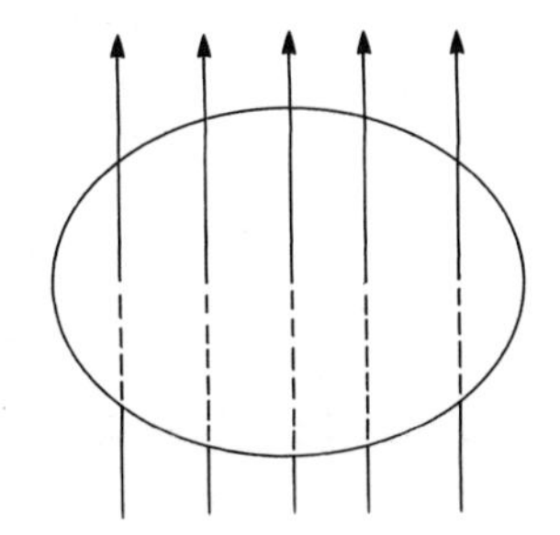

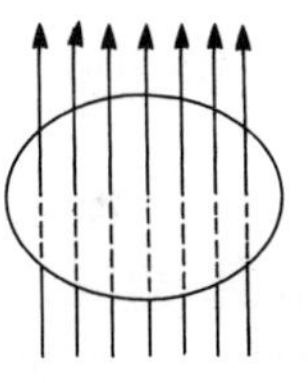

Fig. 15.7. The magnetic field strengths in stellar material can be increased if the material contracts.

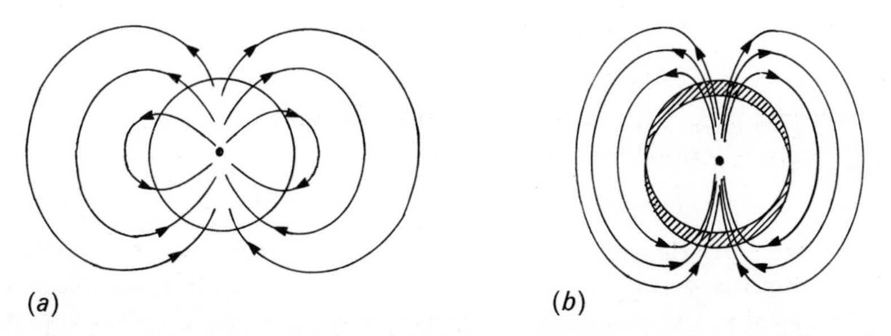

Fig. 15.8. If the equatorial material of the star were to move toward the poles, the magnetic field strength at the poles would increase. It would decrease at the equator. Figure(*a*) shows the original star and its field and (*b*) the shape and the field of the star after the material has moved from the equator to the poles.

peculiar abundances in the photospheres of the peculiar A stars, though there are still some problems if one wants to explain the observed abundances in detail.

15.3 Diffusion in stars

Generally, we know diffusion as a slow process of moving a substance to another place by molecular motion. Light particles can diffuse through a membrane into a mixture with heavier particles, or a gas of one kind can mix with a large volume of a different gas by diffusion which actually means by thermal motion. When we think about a star we usually think about a well-mixed gas ball. In the case of the Ap stars we do not actually think about mixing but rather about demixing or sedimentation. In nature that will never happen without an external force. In stars, the only forces appear to be gas pressure and gravitation. Under the influence of gravitation, we find a stratification of the gas pressure. On a high mountain, the gas pressure is much lower than at sea level. We can calculate how the gas pressure changes with height in the atmosphere if we consider the equilibrium between pressure and gravitation. At each level the weight of the overlying material has to be balanced by the gas pressure P_g, (see Fig. 15.9). At the height h this means

$$P_g(h) = \text{mass}\,(h) \cdot g \tag{15.2}$$

where mass (h) is the mass of the material above the height h, and g is the gravitational acceleration. At the height $h + \Delta h$ we must have accordingly

$$P_g(h + \Delta h) = \text{mass}\,(h + \Delta h) \cdot g \tag{15.3}$$

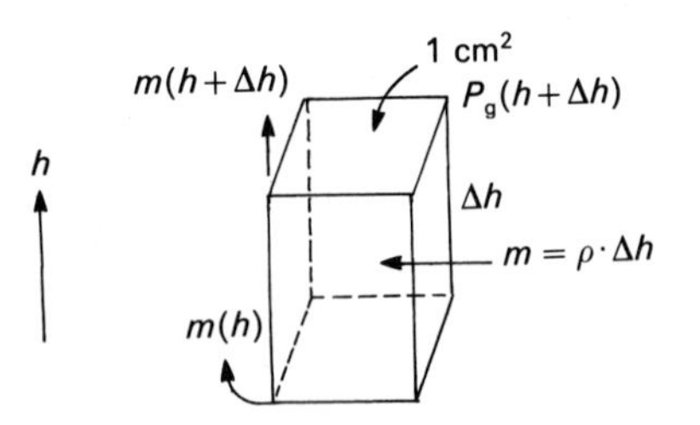

Fig. 15.9. The pressure at the height h has to balance the weight of the overlying material, which means it has to balance the weight of the mass column of $1\,\text{cm}^2$ cross-section above the height h, which is $m(h)$. The pressure at the height $h + \Delta h$ must balance the mass column above $h + \Delta h$, which is $m(h + \Delta h)$. The difference in weight of the two mass columns is given by the weight of the mass in the volume element of $1\,\text{cm}^2$ cross-section and height Δh. This mass is given by $\rho \cdot \Delta h$.

where Δh is the height of the volume element under consideration. The mass (h) is larger than the mass $(h+\Delta h)$ by the mass contained in the volume element. If this volume element has a cross-section of 1 cm^2 and if the density of the material is ρ then the mass in this volume element is given by

$$\text{mass}\,(h) - \text{mass}\,(h+\Delta h) = \rho\cdot\Delta h \tag{15.4}$$

We can now express the gas pressure at the height $h+\Delta h$ by means of the first terms of a Taylor expansion around the height h and obtain

$$P_{\rm g}(h+\Delta h) = P_{\rm g}(h) + \frac{\mathrm{d}P_{\rm g}}{\mathrm{d}h}\Delta h = \text{mass}(h+\Delta h)\cdot g \tag{15.5}$$

and in the same way we find

$$\text{mass}(h+\Delta h) = \text{mass}(h) + \frac{\mathrm{d\,mass}}{\mathrm{d}h}\Delta h = \text{mass}(h) + \rho\Delta h \tag{15.4a}$$

Subtracting (15.2) from (15.5), making use of (15.4a) we derive

$$\frac{\mathrm{d}P_{\rm g}}{\mathrm{d}h}\Delta h = -\rho\cdot g\,\Delta h \quad \text{or} \quad \frac{\mathrm{d}P_{\rm g}}{\mathrm{d}h} = -\rho\cdot g = -\frac{P_{\rm g}\mu}{R_{\rm g}T}g, \tag{15.6}$$

if we replace $\rho = (P_{\rm g}\cdot\mu/R_{\rm g}T)$, where μ = molecular weight of the gas, $R_{\rm g}$ = gas constant = 8.315×10^7 [cgs].

Equation (15.6) tells us that $P_{\rm g}$ decreases with increasing height, and decreases faster if μ is larger. We can see this even better if we integrate (15.6). Division by $P_{\rm g}$ yields

$$\frac{1}{P_{\rm g}}\frac{\mathrm{d}P_{\rm g}}{\mathrm{d}h} = \frac{\mathrm{d}\ln P_{\rm g}}{\mathrm{d}h} = -\frac{\mu g}{R_{\rm g}T}. \tag{15.7}$$

If the right side is a constant, which means for an isothermal atmosphere, we can easily integrate both sides of this equation over dh between the limits $h = h_0$ and $h = h$ and find

$$\ln P_{\rm g}\Big]_{h_0}^{h} = -\frac{\mu g}{R_{\rm g}T}\cdot h\Big]_{h_0}^{h}, \quad \text{since} \quad \int_{h_0}^{h}\frac{\mathrm{d}\ln P_{\rm g}}{\mathrm{d}h}\mathrm{d}g = \ln Pg\Big]_{h_0}^{h}, \tag{15.8}$$

and

$$\int_{h_0}^{h}\mathrm{d}h = h - h_0.$$

If we insert these limiting values, we find

$$\ln P_{\rm g}(h) - \ln P_{\rm g}(h_0) = -\frac{\mu g}{R_{\rm g}T}(h-h_0) = \ln\frac{P_{\rm g}(h)}{P_{\rm g}(h_0)}. \tag{15.9}$$

Taking the exponential yields

$$\frac{P_g(h)}{P_g(h_0)} = e^{-(\mu g/R_g T)(h-h_0)} \quad \text{or} \quad P_g(h) = P_g(h_0) \cdot e^{-(h-h_0)/H}, \qquad (15.10)$$

where we have abbreviated $(R_g T/\mu g)$ by H, which is called the scale height. If $(h - h_0) = H$, then the exponent in the exponential is -1. The pressure decreases by a factor e over one scale height. The scale height H is smaller if μ is larger. Heavy particles form an atmosphere of smaller height than lightweight particles (see Fig. 15.10).

It turns out that in a mixture, each group of particles tries to set up its own equilibrium. If you make an emulsion of heavier particles in water, the heavy particles will sink. If you have a mixture of air with N_2 ($\mu = 28$), CO_2 ($\mu = 44$), and H_2O($\mu = 18$), then H_2O will become more abundant in the higher atmospheric layers, provided there are no atmospheric currents or winds which keep the atmosphere well-mixed.

The stellar material consists of H, He, and heavier particles, and very light electrons. If there were no mixing, the electrons would like to float on top. This would, however, generate a charge separation, creating an electric field which prevents further charge separation (see Fig. 15.11). The electrons and ions are kept together by the electric forces. But there is no force which inhibits the falling of the still heavier ions with respect to the protons. The helium atoms and ions may well fall with respect to the hydrogen atoms and ions, but the density in stars below the surface is so large that there are very many collisions which prevent the particles from falling or diffusing downward fast. It would take more than the lifetime of the stars to produce any measurable separation over the radius of the star, even for the still heavier ions. This was already discussed by Eddington (1926).

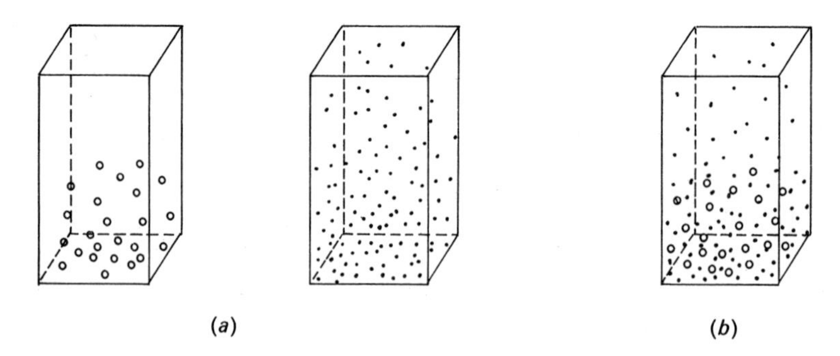

Fig. 15.10. In an atmosphere with heavy particles (*a*) the pressure decreases faster with height than in an atmosphere with light particles. In an atmosphere consisting of heavy and light particles (*b*), the heavy particles stay close to the bottom, the light ones extend to greater heights.

To see deficiencies at the surface does not, however, take as long as diffusion to the center. The layers which we see are only a few hundred km thick. Diffusion can go rather fast in these layers where 'fast' means about 10^5 years, provided there is no mixing mechanism. If measured line widths indicate turbulent motions, then diffusion cannot work.

Also, we are talking about Ap stars, which show an enrichment, not a depletion, of heavy elements like the rare earths and Si, Cr, and Eu at the surface. Why are these elements enriched?

There is another diffusion process which opposes the gravitational settling. This process is called radiative diffusion, and the force causing this diffusion is radiation pressure.

In a star, energy is transported outwards from the inside by radiation. This means that there is more radiation coming from below than from above (Fig. 15.12), otherwise there would be no net flux to the surface and to the outside (see Fig. 15.13(*a*)). The atoms in the atmosphere absorb and re-emit the photons. In each absorption process, the photon transmits some momentum to the absorbing atom or ion. The photon will generally be re-emitted in another direction (see Fig. 15.13(*b*)). In fact, there is no preferred direction for the re-emission. Therefore, in the re-emission the net effect of all the pushes is zero. But this is not the case for the absorption because there is more radiation coming from below than from above, so there is a net push to the surface. The atoms and ions which absorb and re-emit most, receive the biggest push. The rare earth elements have a very complicated energy-level

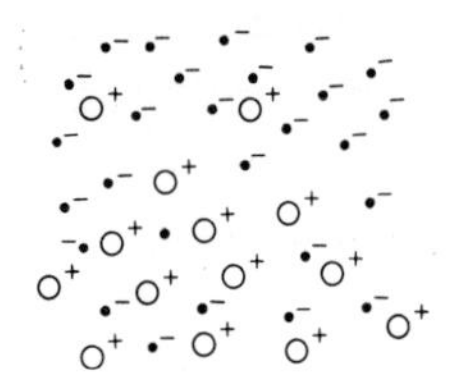

Fig. 15.11. In an atmosphere with free electrons and protons, the electrons would like to diffuse to the top. This would give a charge separation, yielding an electric field, which prevents further separation.

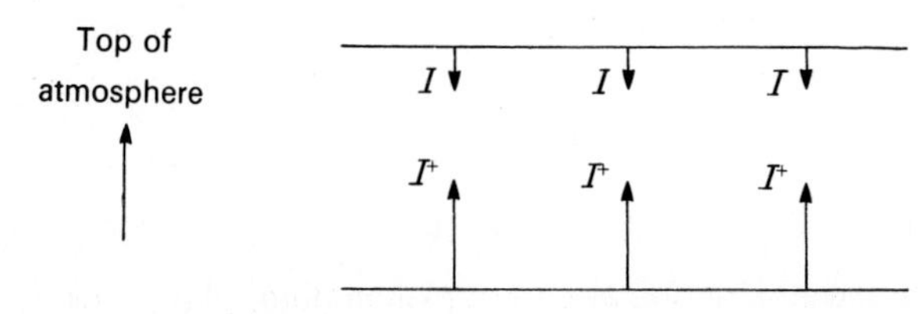

Fig. 15.12. In order to transport radiative energy from the inside of the star to the surface, more radiation has to flow in the outward than in the inward direction. I^- is the downward going intensity, and I^+ the outgoing intensity.

diagram; they have, therefore, very many lines in the spectral region where there is a large flux coming out of the star. Ca^+ has very few lines because it has only one electron in its outer shell. It has more lines in the far ultraviolet but there are no photons to absorb. Therefore, the Ca^+ can sink while the rare earths do not.

Also, He and He^+ have only a few lines because they have only one or two electrons and will, therefore, tend to sink. Helium appears to be depleted in the peculiar A stars. The radiative diffusion can, indeed, be different in stars with different effective temperatures because the degree of ionization changes and the radiation field changes with T_{eff}.

A remaining problem is the convection zone below the surface of the Ap stars which is not important otherwise, but which could keep the top layers well-mixed. Probably the magnetic field is strong enough to prevent these motions. In fact, this is supposed to be the reason why diffusion works in magnetic stars but not in non-magnetic A stars.

Diffusion of charged particles is difficult in a direction perpendicular to the magnetic field lines, but easy along the direction of the field lines. We may, therefore, expect a larger enrichment of the upward diffusing elements in both polar regions. This explains the different element abundances on different parts of the magnetic stars.

15.4 Metallic line stars, or Am stars

The most abundant peculiar stars are called metallic line stars. About one-third of all the late A or early F stars belong to this group. As their name says, the lines of most of the heavy elements are too strong in comparison with their hydrogen line strength. On the other hand, the usually very strong lines of the calcium ion Ca^+, the lines at 3933 and

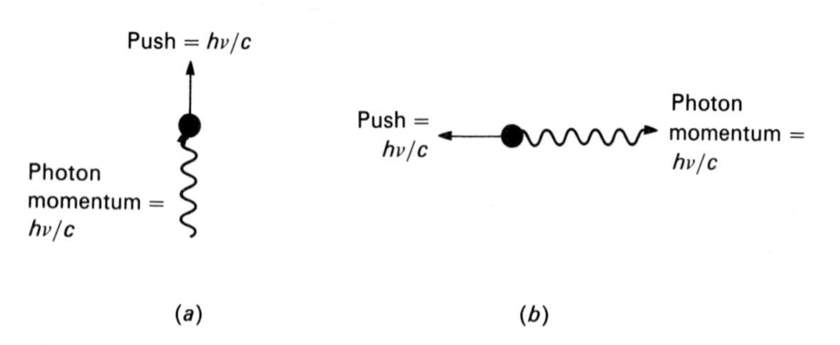

Fig. 15.13. When the atom or ion absorbs a photon from below, it receives a push to the surface (*a*). When the photon is re-emitted in another direction, the atom or ion receives a push in a direction different from that during the absorption (*b*).

3968 Å, the H and K lines, are too weak in these stars (see Fig. 15.14, where we compare a spectrum of an Am star with those of normal A and F stars). In Fig. 15.15 we show schematically for 'normal' stars the dependence of the hydrogen line strengths on the spectral types, the general dependence of the metallic line strength on the spectral types, and also the dependence of the H and K line strengths on the spectral types. While the strengths of the metallic lines in general increase for cooler stars, the H and K line strengths decrease

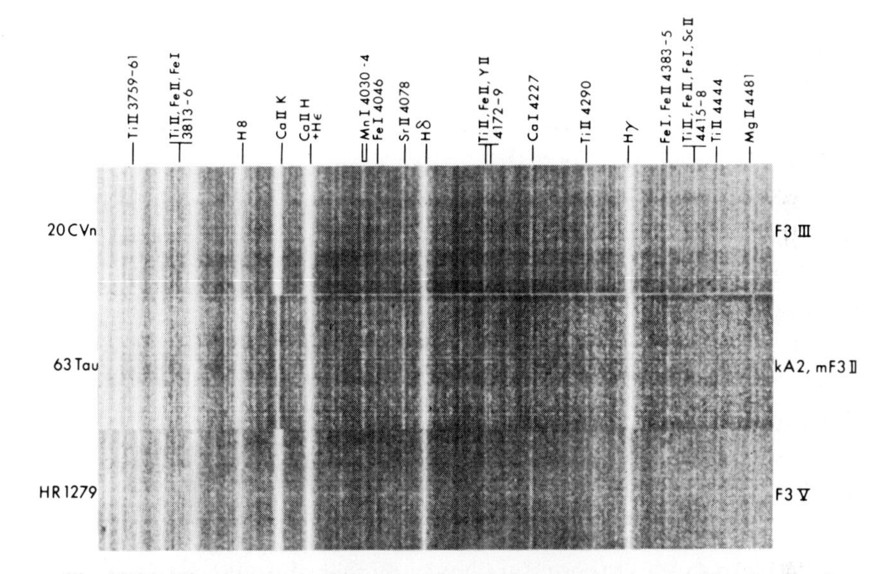

Fig. 15.14. The spectrum of the Am star, 63 Tau, is compared with spectra of normal A and F stars. In the metallic line star the metallic lines look like those of an F star, but the hydrogen lines and the calcium lines look like those of A stars. (From Morgan *et al.* 1978.)

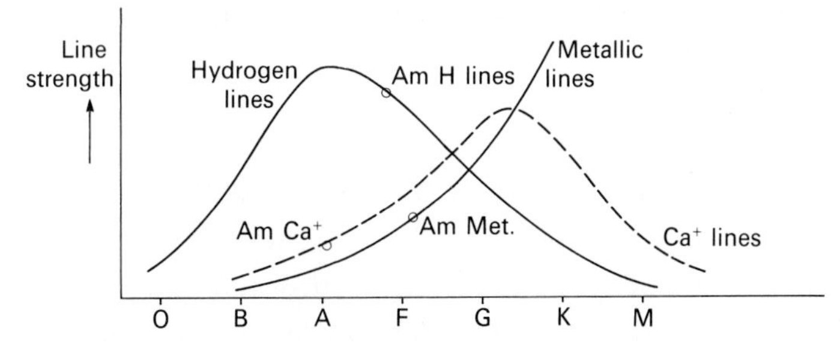

Fig. 15.15. The dependence of the hydrogen, calcium and other 'metallic' line strengths on spectral types for 'normal' stars is shown schematically. For the spectral type indicated by the H lines in an Am star the metallic lines are too strong, or for spectral type indicated by the metallic lines the hydrogen lines are too strong. The Ca^+ K lines are too weak for any of the other spectral types.

for lower temperatures because of the decreasing degree of ionization. In the cooler stars, the lines of the neutral Ca become stronger. If in the metallic line stars the Ca^+ lines, also called the Ca II lines, are too weak, then the spectral type derived from the strengths of these lines will be too early, and we would think the stars are rather hot. On the other hand, if we look at the Fe lines, which appear too strong, we would think the stars are cooler than indicated by the hydrogen lines. If we determine their spectral type according to the metallic line strengths, they would be called late A or early F stars. However, if we classify them according to the Ca II H and K lines at 3933 and 3968 Å, they would be classified as early A stars, i.e., A0 to A3. The strengths of the Hydrogen lines would indicate a spectral type, say, around A5, somewhere intermediate between the metallic line and the Ca II line spectral types.

Which spectral type tells the correct temperature?

We discussed previously that the line strength is determined by the ratio $\kappa_{(\text{line})}/\kappa_{(\text{cont})}$. For A stars, the continuum absorption is due to hydrogen; the hydrogen *line* absorption is also due to hydrogen. It is very difficult to envisage any mechanism which could reduce the hydrogen line strengths except a lower density, which would require the stars to be giants. But many of the Am stars are members of clusters: the Hyades cluster, for instance. Their brightness shows that they are not giants even though the ionization equilibrium of iron indicates a somewhat reduced pressure in the atmospheres of these stars.

The simplest way to explain the peculiar line strengths of the metallic lines would be to say that these stars have a higher abundance of the metals, but they must then have a lower abundance of Ca. It turns out that the scandium lines are also weak, as are sometimes, the carbon lines and those of some other elements. The lines of the rare earth elements are rather strong, compared to the Fe lines. We then must ask: what can cause these peculiar element abundances?

15.4.1 The binary nature of Am stars

These Am stars have some other peculiarities. When a careful study of their radial velocities was made, it turned out that they all have periodically variable radial velocities, which means they are all binaries, and all rather close binaries, with periods of the order of days to weeks. (For 'normal' stars, about 30% of all are binaries). What does that have to do with their peculiar abundances? How short can the periods of binaries be without the stars touching each other? For the stars not to touch, the distance *d* must be larger than $\sim 2R^*$. For Am stars, $R \sim 2R_\odot$. This means

$$d \gtrsim 4R_\odot.$$

$$1R_\odot \sim \frac{1\text{ au}}{200} \quad \text{so} \quad d > \tfrac{1}{50}\cdot\text{au}.$$

If we measure all quantities in solar units, we know that Kepler's third law reads

$$\frac{d^3}{p^2} = M_1 + M_2.$$

where p is the period, and M_1 and M_2 are the masses of the two stars. With $M_1 + M_2 \sim 3M_\odot$ and $d = 1/50$ au, we find for the period in years,

$$p^2 \geqslant (\tfrac{1}{50})^3 \cdot \tfrac{1}{3} \sim \tfrac{1}{3} \cdot 10^{-3} \cdot \tfrac{1}{125} \sim \tfrac{1}{4} \cdot 10^{-5} \text{ years}$$

or $P \geqslant 1.5 \times 10^{-3}$ years, or roughly $P \geqslant 1$ day.

The Am stars have periods longer than that, but not much longer. They must have strong tidal interaction, but are *not contact* binaries.

The Am stars are also peculiar in another respect, as we will discuss in the next section.

15.4.2 Rotational velocities of Am stars

As we saw in Chapter 12 we can measure the rotational velocities of stars from the widths of their spectral lines. Rotational velocities for many stars have been determined in that way. In Figs 13.5 and 13.6 we have shown the dependence of the average rotational velocities on the spectral types. The rotational velocities decrease rapidly for stars cooler than spectral type F5. It is believed that this decrease of rotational velocity is due to the development of a stronger stellar wind, like the solar wind, which causes the braking of rotation.

The Am stars occur in the region of spectral classes where the rotational velocities are generally still high, but the Am stars have much smaller rotational velocities than the normal stars.

What does all this have to do with the stronger metallic lines and with the apparent over-abundance of heavy elements?

Most astronomers believe that the peculiar abundances observed in the Am stars are also due to diffusion processes even though there are still some severe problems with that explanation: if Ca is depleted due to the sinking of Ca^+ in the layer where calcium is mainly ionized once, why is not Mg also depleted? Its atomic structure is very similar to Ca, yet the Mg abundance is normal in the Am stars.

The other question is, of course, why should diffusion work in Am stars and not in normal stars? Why are all the Am stars in binaries and why are they slow rotators? The reason is again that diffusion can only work if there are no mixing processes which prevent it. Normal stars of the same temperature as the Am stars would still be rapid rotators. Since the studies of Eddington, it is well known that rapidly rotating stars cannot be in a static equilibrium, but that slow currents, the so-called Eddington currents, develop which carry material upward in the polar regions and downward in the equatorial region. These currents can keep the star well-mixed. Therefore, diffusion can only work for slowly rotating stars. What makes the Am stars rotate slowly? Their binary nature – they are all members of rather close binaries, for which tidal interaction is very important. Tidal interaction slows down rotation until the rotational period equals the orbital period.* For an orbital period of one week, this would mean an equatorial rotational velocity of about 15 km/s for the Am stars, which would be considered a slow rotation. The binary nature then causes the slow rotation which in turn may permit diffusion to work.

15.5 Barium stars

There is another group of stars with peculiar element abundances, called barium stars. These have gained interest during recent years even though they are relatively rare stars. They are certainly not made out of barium, but show unusually strong lines of barium and strontium (see Fig. 15.16). In Table 15.1, we list the element abundances as determined for some of these stars. They show a general increase of abundances for elements which are made by the so-called slow neutron capture process. This means that the atomic nuclei of these elements were originally formed by adding neutrons to the atomic nuclei in a process called neutron capture. Neutron rich atomic nuclei become unstable to the so-called β decay, which means they like to emit an electron thereby transforming a neutron into a proton. Very neutron rich atomic nuclei cannot bind any more neutrons, they must wait until one of the captured neutrons has emitted an electron before it can capture another neutron. Due to the decay the number of surplus neutrons is reduced and an additional neutron can be bound. The transformation of a neutron into a proton can occur for different numbers of surplus neutrons,

* Our moon rotates once in 27 days, which means during its orbital period. We therefore always see the same side of the moon.

depending on the available number of neutrons to be captured. If neutrons are added very rapidly in a very neutron rich environment, then there is no time for the less neutron rich nucleus to decay before the next neutron is added. We call this the 'fast' neutron capture process. If fewer neutrons are around then neutron captures are less frequent and the less neutron rich nucleus has time to emit an electron before capturing the next neutron. This is called the 'slow' neutron capture process. The resulting atomic nuclei which are formed by these different neutron capture processes are different. From the observed abundances we can thus see which of the processes has taken place. For the barium stars we see the products of the slow neutron

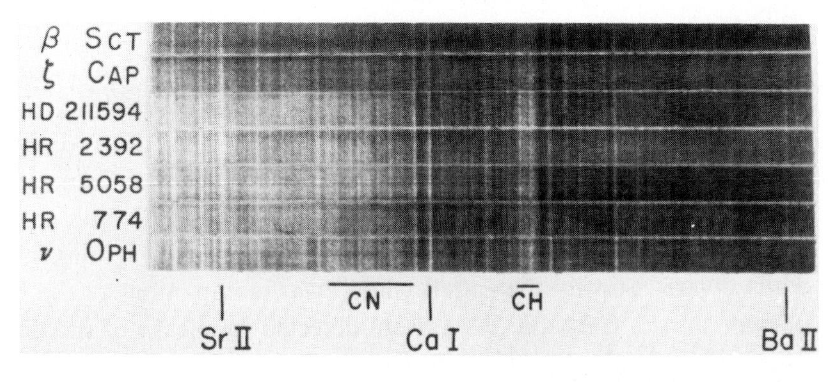

Fig. 15.16. The spectra of five barium stars are compared with the spectra of the 'normal' stars β Scuti and ν Ophiuchi. In the Ba stars the Ba II line at 4554 Å is much stronger than in the normal stars and also the Sr II line at 4078 Å is stronger. (From Bidelman & Keenan 1951.)

Table 15.1 *Abundances*[1] *of slow neutron capture elements in two barium stars and in the sun.*

Element	log N Solar	HR 774	HD 44896	$\bar{\Delta}^2$
Sr	2.94	4.22	3.79	1.06
Y	2.23	3.34	3.44	1.16
Zr	2.59	3.59	3.45	0.97
Ba	2.20	3.28	3.07	0.97
La	1.22	2.30	2.27	1.06
Eu	0.55	1.32	1.07	0.65
Gd	1.09	2.04	1.80	0.83
H	12.0	12.0	12.0	0.0

1. The numbers given are the logarithms of the number of atoms plus ions in the amount of material which contains 10^{12} hydrogen atoms and ions.
2. $\bar{\Delta} = \log N_{\odot} - \log N$ (Ba star).

capture process. This generally means that between successive neutron captures there was a timespan of roughly 10^4 years. There must have been a weak neutron source present for at least that length of time in order to make the slow neutron capture process work. Such slow neutron capture processes are believed to be possible only in the interior of stars which are in very advanced states of their evolution. At that time they are supposed to be very bright red giants, as we shall see in Volume 3. The barium stars are not that bright, though they are red giants. It has been a puzzle for many years how these stars could have manufactured these slow neutron capture elements and, in addition, if they did so, how they could have mixed the processed material to the surface where we can observe them. There is presently no deep mixing.

The problem appeared in a different light when McClure, Fletcher, and Nemec (1983) discovered that all the barium stars showed radial velocity variations indicative of their binary nature. From the radial velocity variations of some of the barium stars, they concluded that the masses of the companions were of the order of one solar mass. Since the companion spectra could not be detected, they speculated that the companions might be white dwarfs. Shortly thereafter, white dwarf companions of the brightest barium stars, ζ Cap and ζ Cyg, were detected by means of the ultraviolet observations with the IUE (International Ultraviolet Explorer) satellite. ζ Cyg is a so-called mild barium star which shows only a small enhancement of the barium line intensity. An additional white dwarf companion was detected for the mild barium star ξ^1 Cet. For other nearby barium stars, the presence of a white dwarf companion appears probable because emission lines indicative of a rather hot gas are seen. No main sequence companions with masses comparable to the masses of the barium stars, about two solar masses, could be detected, though they should be visible in the ultraviolet if present. This supports the suspicion that all the barium star companions are, indeed, white dwarfs. If the companions were mainly normal stars, the chances of finding stars of almost the same mass as the barium stars would be rather large since statistics show that about half of all binaries have two components with about equal masses, to within 50%. Since we do not find companions which are normal stars we conclude that all barium star companions are white dwarfs.

As we shall see in Volume 3, white dwarfs are believed to be the final death stages of stellar evolution. They supposedly form when a star blows off its outer envelop at the state of stellar evolution when it is a very luminous red giant. The exact cause of this mass loss is presently still unknown, but we know that it must happen. The reasons will be discussed in our stellar evolution discussions.

If, indeed, all barium stars have white dwarf companions, then this must be a necessary condition for the formation of these stars. The presence of the white dwarf must be essential to the peculiar element abundances. It seems likely that the slow neutron capture process did actually take place in the interior of the progenitor star of the present white dwarf, which must have been a very luminous red giant before it lost its envelope and became a white dwarf. The barium star probably accreted a major part of the envelope of its companion when it was blown off. We probably see the material manufactured in the interior of the white dwarf progenitor displayed now on the surface of the present barium star. There are still many studies needed before we can be sure that this is actually what happened, but these observations open the exciting possibility that we can actually study, on the surface of the barium star, what happened in the interior of the white dwarf progenitor.

15.6 T Tauri stars

The so-called T Tauri stars are named after their prototype. The T in front of the constellation indicates that the stars are variable, they are in fact irregularly variable. This type of star is defined by the peculiar spectra which show strong emission lines of hydrogen and of the Ca^+, H and K lines at 3933 Å and 3968 Å (see Fig. 15.17). They frequently also show emission

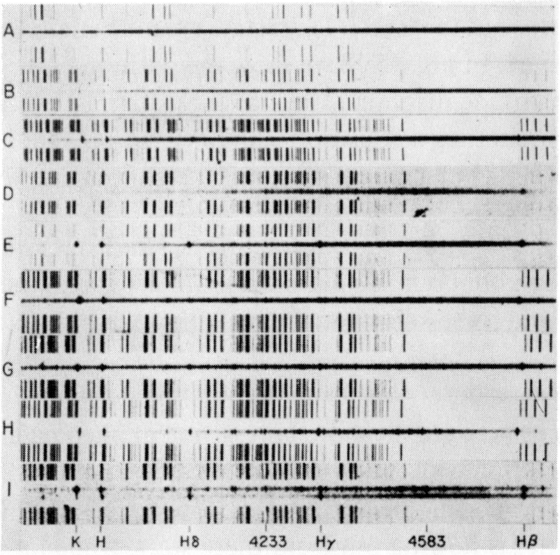

Fig. 15.17. Negative T Tauri spectra (the narrow horizontal lines) are shown for different T Tauri stars. The strong emission lines of the Ca^+ ion, the H and K lines, are seen (dark blobs on the left). On both sides of the stellar spectra are laboratory spectra, such that wavelengths can be measured for line identification. (From Joy 1960.)

lines of other elements like iron or sulfur. The particular lines of these latter elements are not usually seen in normal stellar spectra but are lines which are generally seen in objects of very low density like gaseous nebulae. They are the so-called forbidden lines indicating low density material around these stars.

As judged from the absorption line spectra also seen in these T Tauri stars they are of spectral types F to M which means they have effective temperatures between 7000 and 3500 K. As discovered by ultraviolet obserations by means of the IUE satellite these stars show very strong chromospheric and transition layer emission lines much stronger than those of the sun (see Chapter 18).

These T Tauri stars are especially interesting because they appear to be very young stars as demonstrated by the fact that they appear only in regions of the sky where we also see hot O and B stars, which we know cannot live longer than about 10^6 years. The T Tauri stars always appear in dense interstellar gas clouds which we think are the birth places for stars. If such a gas cloud is large and dense enough the gravitational force can pull it together to form a star. The association of these stars with young O and B main sequence stars whose absolute magnitudes we know and for which we can therefore determine the distances (see Chapter 5) makes it possible for us to also get approximate distances to the T Tauri stars, though the absorption by the gas in the cloud surrounding the star introduces some uncertainty, (see Chapter 19). It turns out that the T Tauri stars are much more luminous than main sequence stars. In Fig. 15.18 we show the position of the T Tauri stars in the color magnitude diagram. The stars are still much larger than main sequence stars. From their association with the young O and B stars we know they must be very young. They must still be contracting finally to become main sequence stars.

From the study of these young, still contracting T Tauri stars we can get interesting information about our sun at an evolutionary stage when it was still contracting and when the solar system, including the planets and the earth was forming.

A more detailed discussion of star formation will be given in Volume 3. What is of interest to us here are the more or less violent phenomena which seem to occur in connection with the birth of a star. For many T Tauri stars we see for instance P Cygni line profiles (see Chapter 17, Fig. 17.9) indicating that material is flowing out from these stars, while for a newly forming star we would naively have expected that matter is falling inward. It appears that newly formed stars develop very thick chromospheres, coronae and stellar winds, though X-ray emission from the coronae is not observed.

We do not know whether the coronae do not become very hot because of the large energy loss in the strong winds or whether the material in the dense gas cloud in which the star is embedded absorbs all the X-rays.

The T Tauri stars also show collimated, high velocity outflows with velocities of 300–400 km/s, so-called jets (see Fig. 15.19) going in opposite directions, so-called bipolar outflows, which in their extension often show Herbig–Haro objects, named after the astronomers who first discovered and discussed them. These Herbig–Haro objects change their appearance in the course of a few years, (see Fig. 15.20). It was therefore at some time suggested that they were actually newly formed stars and that we could witness directly the birth of new stars. The masses of these objects are, however, much too small. The amazing thing about the Herbig–Haro objects is their fast motion away from the central source, always believed to be a T Tauri star. The

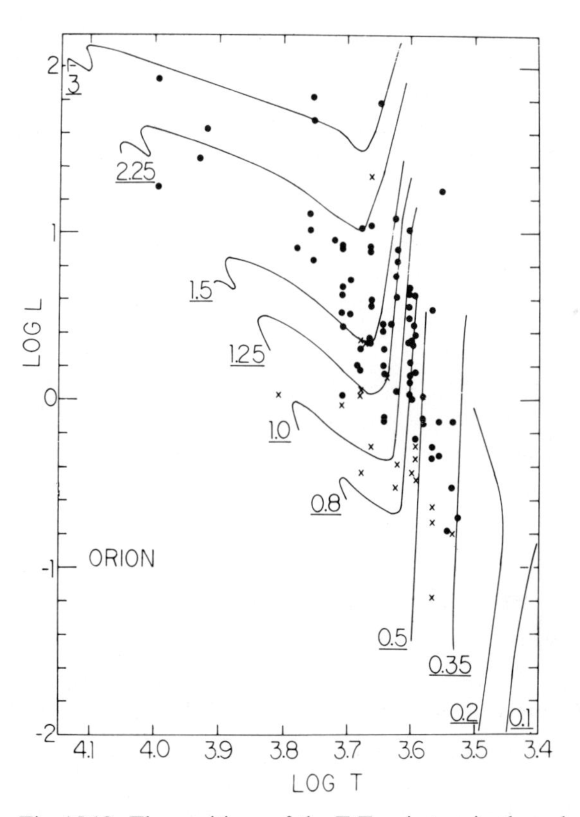

Fig. 15.18. The positions of the T Tauri stars in the color magnitude diagram are shown. They all lie above the main sequence, which is the boundary with the numbers on the left hand side. They must, therefore, be larger than main sequence stars. The solid lines are evolutionary tracks (see Volume 3) for stars of different masses as given by the numbers (in units of solar masses). (From Cohen & Kuhi 1979.)

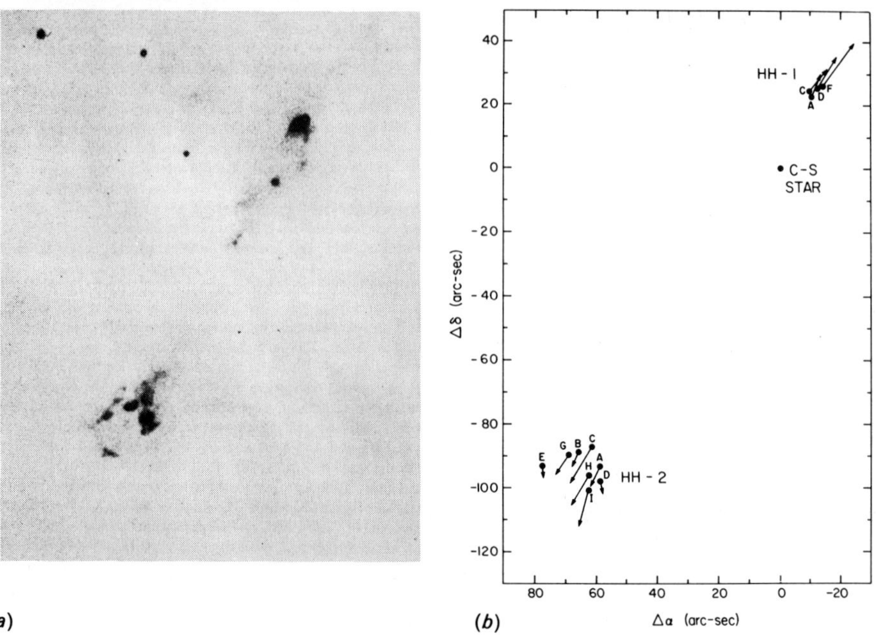

Fig. 15.19. A photograph of the Herbig–Haro objects HH 1 and HH 2 is shown on the left (*a*). (This is again a negative, the dark spots are bright.) Their proper motions (corresponding roughly to 250 km/s) point radially away from a source which is about halfway in between HH 1 and HH 2. It can only be seen in the infrared because all the other light is absorbed by the interstellar dust (see Chapter 19) around it. In the right-hand picture (*b*) the directions of proper motions are shown. The lengths of the arrows are proportional to the measured velocities. (From Schwartz 1983. Originals from Herbig & Jones 1981.)

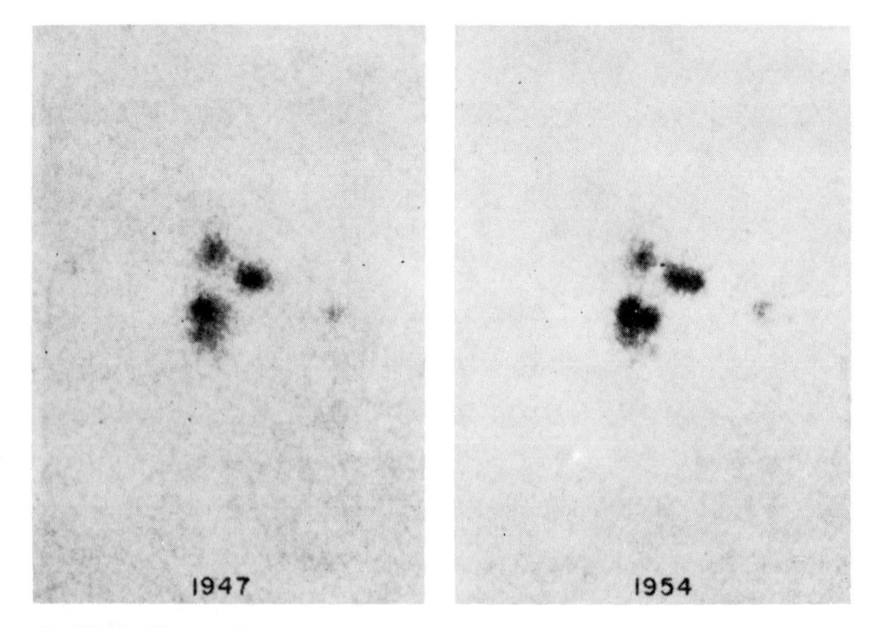

Fig. 15.20. Changes in the appearance of the Herbig–Haro object no. 2 are seen. In 1947 only three major blobs are seen; in 1954 five can be recognized. Two new ones have appeared in the course of seven years. (From Herbig 1969.)

kinetic energy emitted in these clumps of matter corresponds to the energy emitted by the sun during a whole year. It is not at all clear what the energy source is for these jets.

If the original gas cloud from which the star formed, had some angular momentum, we might expect a disk to form in a star forming collapse. We might then understand why a stellar wind can only flow out in directions perpendicular to the plane of the disk and form a jet. It is still surprising that the jet keeps so narrowly focussed, the opening is about 10 degrees, over distances of the order of 1 pc, and the energy source for the jet is still a puzzle.

The spectra of the Herbig–Haro objects can be explained by a spectrum of a shock, forming either when a high density, high velocity gas ball, a so-called interstellar bullet, hits a less dense gas cloud at rest, or forming when a high velocity, low density jet hits a clump of matter.

16

Pulsating stars

16.1 The different types of pulsating stars

The pulsating stars (not to be confused with pulsars, which are rotating not pulsating), are recognized by their periodic changes in brightness accompanied by periodic variations of their radial velocities. We distinguish several classes of pulsating stars. The most frequent type are the δ Cephei stars, which have periods of light variation of a few days to a few weeks. Another very common type is the RR Lyrae star, named after the first-discovered variable of this kind. The RR Lyrae stars vary with periods of the order of half a day. Other types of short period variables are the δ Scuti stars, which are population I variables like the longer period δ Cephei stars. The δ Cephei stars are generally of spectral types F and G, and are supergiants, while the δ Scuti stars are A stars of luminosity class IV to V. The amplitudes of the δ Cephei variables are of the order of one magnitude, i.e., the light output changes by a factor of 2 or 3, while the light variations of the δ Scuti stars are so small that they are hard to detect.

The RR Lyrae stars are frequently found in globular clusters; therefore, they belong to population II. The amplitudes of their light variations are also of the order of one magnitude, and can therefore be recognized fairly easily. The radial velocity amplitudes for the δ Cephei stars are several tens of kilometers per second. Those of RR Lyrae stars are of comparable values, though somewhat smaller. In Fig. 16.1 we reproduce the light and velocity curves of the prototype, namely δ Cephei.

There are also population II star analogues to the δ Cephei stars, called the W Vir stars, again named after the first star discovered of this type.* These

*We should mention here, perhaps, that variable stars are named after the constellation in which they are seen. The first one discovered in the constellation of Virgo would be called A Vir, the second one B Vir. After all the letters have been used, one would continue with AA Vir, AB Vir, etc. Historically it developed though that the first ones discovered were actually designated with letters occurring late in the alphabet.

stars frequently show very broad light maxima. In Fig. 16.2 we show the light curve of the population II Cepheid, W Vir.

Another group of intrinsically variable stars are the long period variables with periods of the order of several years. They are very luminous red giants and are of spectral type M. Their light variations are not quite as regular as those of the other variables discussed so far.

Even less regular are the light variations of the so-called RV Tau stars which, in the HR diagram, are placed between the Cepheids and the long period variables. In Fig. 16.3 we show the positions of all the different kinds of intrinsically variable stars in the HR diagram.

The most frequent types of variables are found in a narrow band cutting diagonally through the HR diagram along the so-called Cepheid instability strip. All stars inside this band in the HR diagram are pulsating stars. In Volume 3 we will explain why.

These pulsating stars actually reveal many properties of the stellar

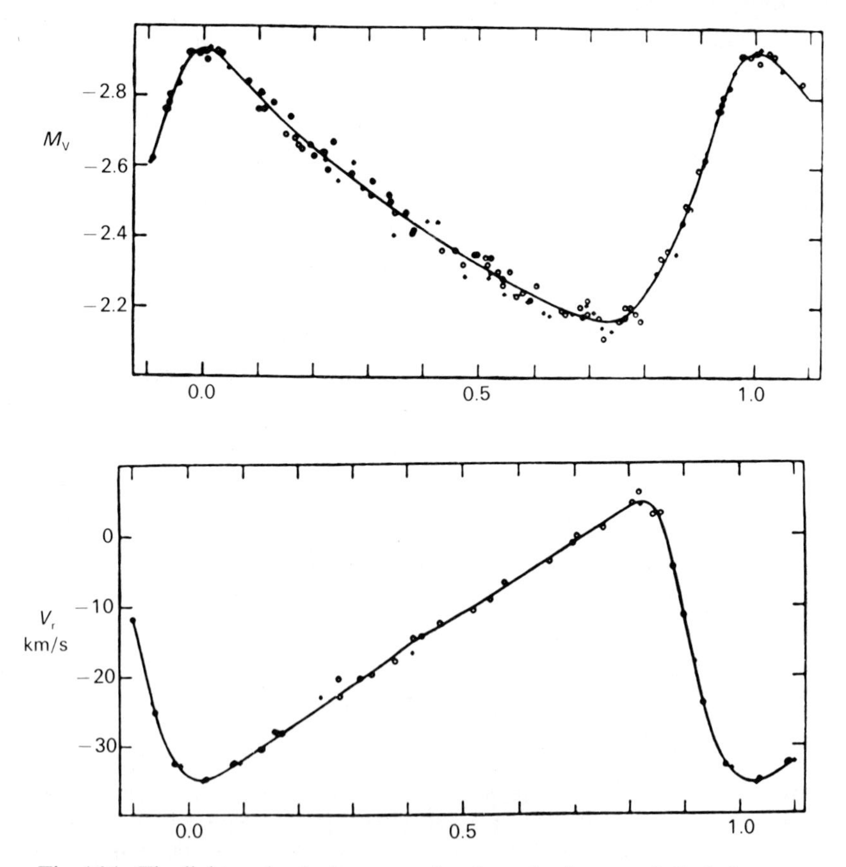

Fig. 16.1. The light and velocity curves for the pulsating star δ Cephei are shown as a function of phase in the pulsational cycle. For this star the period of variation is 5.4 days. (From Cox 1980.)

interiors to us. They serve to check our ideas about the interior structure of the stars and about stellar evolution. They are, therefore, very important to the theory of stellar structure and have been studied extensively.

So far we have mentioned only stars which pulsate in a spherically symmetric way, the surface moving in and out. There are other types of variables which pulsate in a non-spherically symmetric way, i.e., the polar regions may be moving inwards while the equatorial regions move outwards. More complicated patterns are also possible. Spherically symmetric pulsations may be superimposed. The most prominent stars of this kind may be the β Cephei stars. They are B type stars and are not positioned in the Cephei instability strip. The reasons for their pulsational instability is presently unknown.

There is also a group of white dwarfs, the ZZ Ceti stars, which are pulsationally unstable and show a very complicated pattern of pulsation consisting of the superposition of pulsations of many frequencies. They are very helpful for studying the interior structure and chemical composition of these white dwarfs.

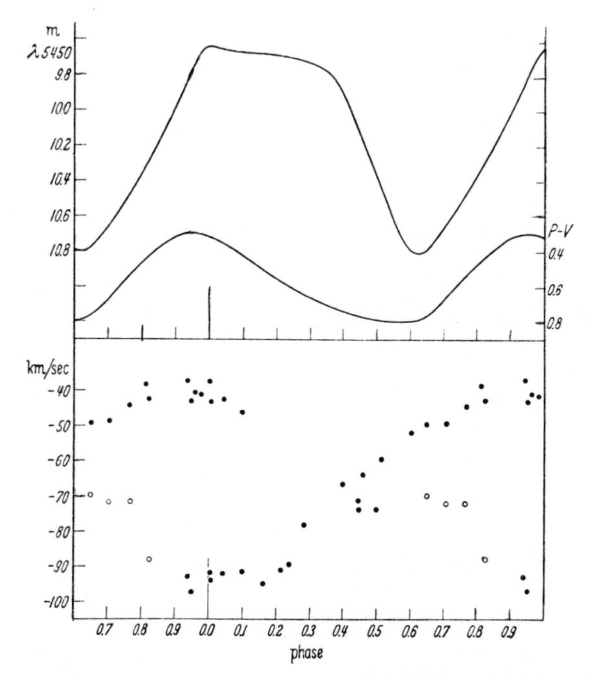

Fig. 16.2. The light curve of the population II Cepheid W Virginis is shown. In comparison with the light curve of δ Cephei it shows a very flat top. Such flat-topped light curves can be well distinguished from those of population I Cepheids, but not all population II Cepheids have flat-topped light curves. The velocity curve is shown at the bottom. The open circles indicate velocities from emission lines. For phases between 0.9 and 0.6 it appears that two absorption line components are seen. (From Ledoux & Walraven 1958.)

In the following sections, we will describe the major types of pulsating stars, the δ Cephei stars and the RR Lyrae stars, in somewhat more detail.

16.2 The δ Cephei stars

Because of their periodic light and radial velocity variations, astronomers first thought that the δ Cephei stars were eclipsing binaries. In Fig. 16.4 we show again the light and velocity variations observed for δ Cephei to be compared with those observed for an eclipsing binary (see Fig. 9.6). The non-sinusoidal velocity curve for δ Cephei would suggest a very elliptical orbit (see Fig. 9.8). The second eclipse would have to be invisible. The stars would have to be very close companions because there is no constant light phase between eclipses. Also, the periods are rather short, only a few days.

When astronomers learned how to determine the luminosities of the stars

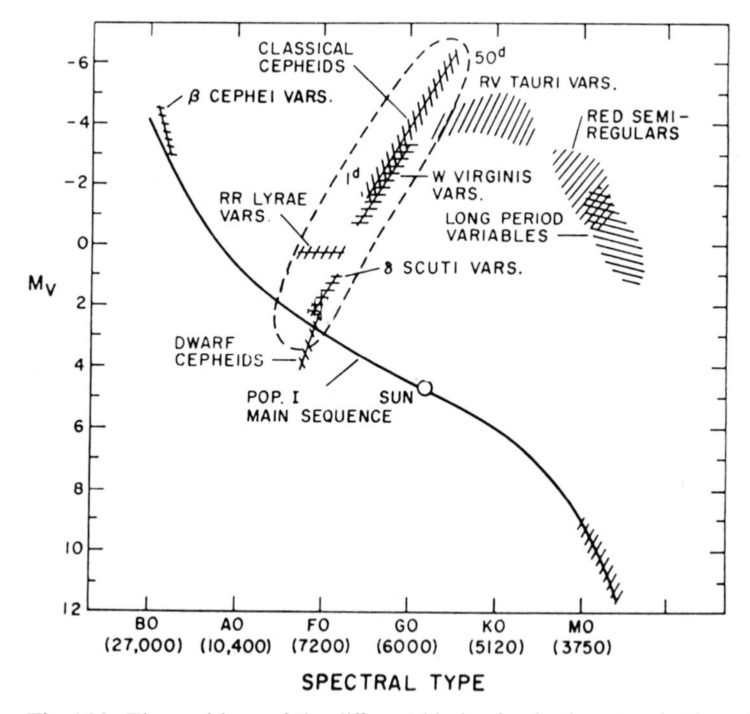

Fig. 16.3. The positions of the different kinds of pulsating stars in the color magnitude diagram are shown. The main groups of pulsating stars fall along a diagonal sequence, the so-called Cepheid instability strip. The long period Mira variables, the irregularly variable RV Tauri stars and, surprisingly, also the β Cephei stars, fall essentially along a constant luminosity strip close to the top of the diagram. The pulsation mechanisms for all these latter stars are not yet understood. (Adapted from Cox 1980).

from their spectral features, it became obvious that the δ Cephei stars are supergiants. This means that the stellar radii are actually larger than were the orbital radii calculated for the stars when interpreting them as binaries. It then became clear that these stars could not be eclipsing binaries, but had to be intrinsically variable stars.

The variable radial velocities indicate that the surface is moving outwards when the light increases and it is falling inwards when we see minimum light. We can directly measure the change in radius $\Delta R(t)$ by calculating

$$\frac{dR}{dt} = v \quad \text{and} \quad \Delta R(t) = \int_{t_0}^{t} v\, dt.$$

In fact, we can also calculate the radius itself if we compare two phases of

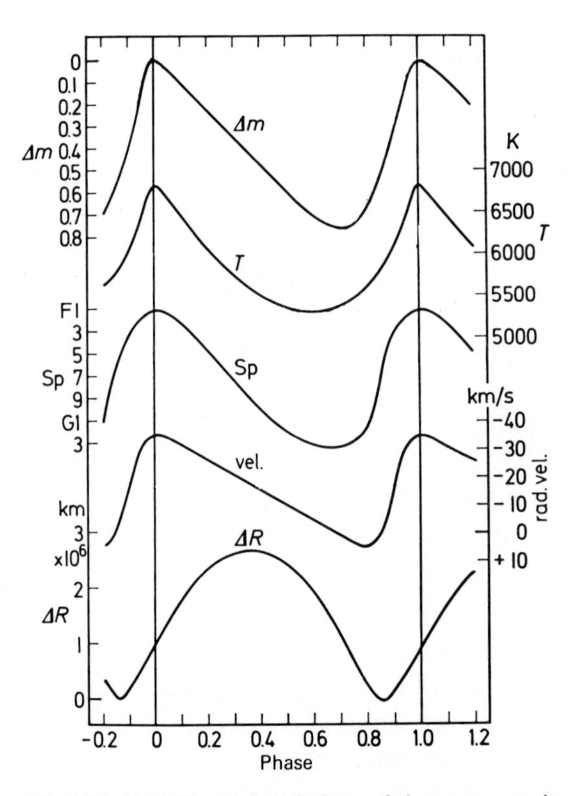

Fig. 16.4. The observed variations of the apparent visual magnitudes, the effective temperatures, the spectral types, and the radial velocities for δ Cephei are shown as functions of phase. At the bottom the changes in the radius are shown, which can be obtained by integrating over the pulsational radial velocities (the stellar velocity has to be subtracted from the observed velocities). It can be seen that the radius is nearly the same for maximum and minimum brightness. (From W. Becker 1950.)

equal temperature. The difference in bolometric magnitude is given by

$$\Delta m_{\text{bol}} = -2.5 \cdot \log \frac{L(t_1)}{L(t_2)} = m_{\text{bol}}(t_1) - m_{\text{bol}}(t_2) \tag{16.1}$$

and

$$m_{\text{bol}}(t_1) - m_{\text{bol}}(t_2) = -2.5 \cdot \log \frac{R^2(t_1)}{R^2(t_2)} = -5 \cdot \log \frac{R(t_1)}{R(t_2)}. \tag{16.2}$$

The Δm_{bol} can be measured; in fact, $\Delta m_{\text{bol}} = \Delta m_{\text{v}}$ since, for equal temperature, the bolometric correction is nearly the same. The change with gravity is minute. We can then calculate the ratio $R(t_1)/R(t_2)$ from (16.2). We can also calculate

$$\Delta R = R(t_2) - R(t_1) = \int_{t_1}^{t_2} v(t)\,\mathrm{d}t. \tag{16.3}$$

Equations (16.2) and (16.3) are two equations for the determination of $R(t_1)$ and $R(t_2)$. These stars change their radii by 5 to 20% during one cycle, i.e., during days or weeks, and their radii are about 50–100 solar radii. The velocities are of the order of 30 km s^{-1}. If the period is 10 days ($\sim 10^6$ s) and if they move outwards during half that time with an average velocity of 15 km s^{-1}, we can estimate ΔR to be

$$\Delta R \sim 1.5 \times 10^6 \cdot 5 \times 10^5 \text{ cm} \sim 7.5 \times 10^{11} \text{ cm} \sim 10\, R_\odot$$

For a radius of 50 $R_\odot$ this means a 20% change. The stars are really blown up like a balloon.

The method of determining the radii of pulsating stars was developed by Baade and Wesselink and is, therefore, called the Baade–Wesselink method.

Using the B − V colors, measured as a function of phase in the pulsational cycle, we can determine the effective temperature also as a function of phase.* In Fig. 16.4 we reproduce the results obtained for δ Cepheii, as given by W. Becker. We find that the star has the highest T_{eff} during maximum light and the lowest T_{eff} during minimum light. This means that the observed light variations are mainly due to temperature changes; in fact, the stellar radii during minimum and maximum light are nearly identical. Maximum outward radial velocity also occurs during the maximum temperature phase. The outward velocity then slowly decreases and the matter falls back onto the star while it cools off. Any theory of stellar pulsations has to explain these phase relations. If the stars were to pulsate in an adiabatic way, we would expect the highest temperature for maximum contraction, which means for

* The phase is defined as the fraction of the period which has elapsed since maximum light.

the smallest radius and the lowest temperature for the largest radius. This is not what is observed.

The lengths of the periods for stars of given luminosity, the shape of the light and velocity curves and the phase relations all depend on the internal structure of the star, as we shall see in Volume 3. By means of these pulsating stars we can therefore check our understanding of the interior structure and the evolution of stars.

16.3 The RR Lyrae stars

Where the Cepheid instability strip cuts through the horizontal branch of the population II HR diagram, we find the RR Lyrae stars. This

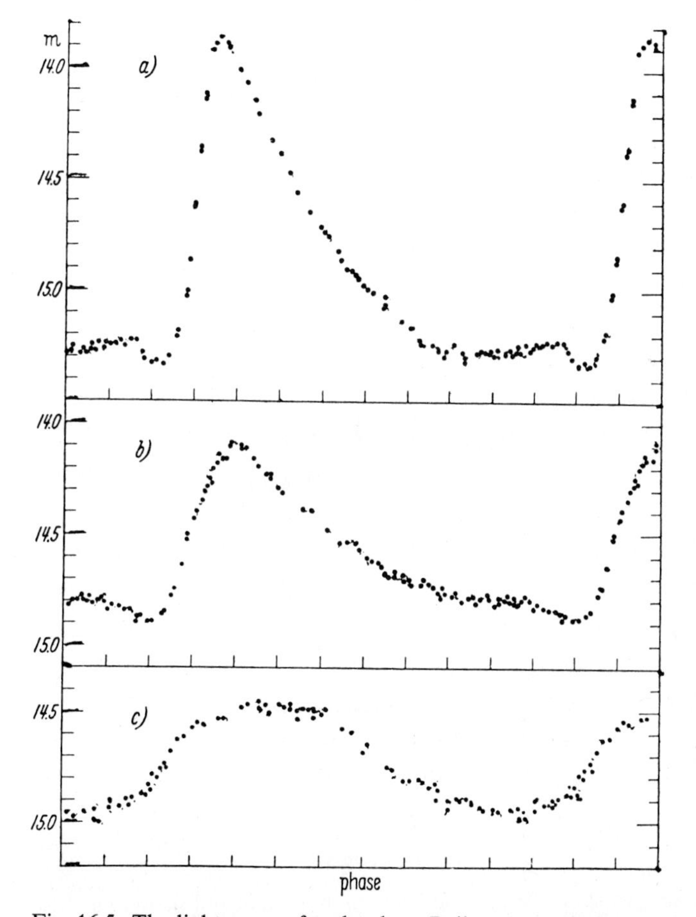

Fig. 16.5. The light curves for the three Bailey types of RR Lyrae stars are shown. Types *a* and *b* only differ in amplitude. The *c* RR Lyraes show a nearly sinusoidal light variation. (From Ledoux & Walraven 1958.)

group of pulsating stars is found in globular clusters as well as in the field. They have periods of around 1/3–1 day.

People distinguish three types of RR Lyrae stars: Bailey types *a*, *b*, and *c*, called so after the astronomer who first made this distinction. Light curves for the three types are shown in Fig. 16.5. While the type *c* stars have almost sinusoidal light curves, the types *a* and *b* have very non-sinusoidal curves. The main difference between types *a* and *b* is the amplitude, and there is really no basic difference between these two types except for the smaller amplitude and the somewhat longer periods for the type *b* stars. But there is a very basic difference between types *a* and *c*.

In Fig. 16.6 we show the relation between amplitude and period for RR Lyrae stars. Again we see that there is a smooth transition between type *a* and type *b* RR Lyraes, while those of type *c* are clearly distinct. The type *c* have the shorter periods. They have all periods less than 1/2 day, while the *a* and *b* types all have periods longer than 1/2 day. It turns out that the type *a* and *b* RR Lyrae pulsate in their fundamental mode, i.e., with the longest possible period, while the type *c* RR Lyraes pulsate in the first overtone.

There are also RR Lyrae stars which have variable amplitudes in light variation, like the star RR Lyrae itself. We can understand this as a beat phenomenon of several periods.

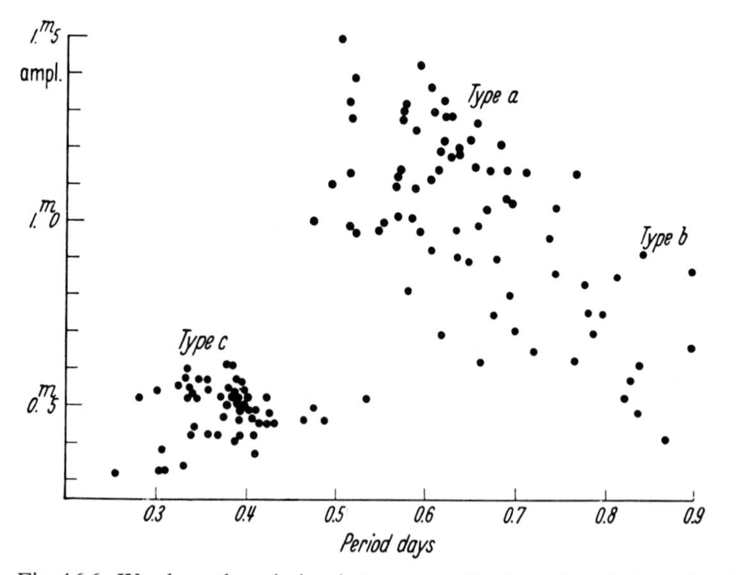

Fig. 16.6. We show the relation between amplitude and period for the three types of RR Lyrae stars. The *c* types have the shortest periods and the lowest amplitudes. They are clearly distinct from the *a* and *b* types. (From Ledoux & Walraven 1958.)

16.4 The period–luminosity relation

The δ Cephei stars are also very important for a very different reason, namely their period–luminosity relation. For the Cepheids found in the nearest external galaxy, the Large Magellanic Cloud, Ms Levitt discovered that the brightest Cepheids had the longest periods. When she plotted the luminosities of the Cepheids as a function of their periods, she found a rather tight correlation. For the Cepheids within the Large Magellanic Cloud we know that they must all be at nearly the same distance because the diameter of this galaxy is much smaller than the distance from us. This period–luminosity relation found for the variables in the Large Magellanic Cloud therefore showed that it is intrinsic to the Cepheids and can be used for other Cepheids as well. Once it is calibrated, which means once we know the absolute magnitude for one Cepheid of a given period, we can, in fact, use this period–luminosity relation to determine the distance to any Cepheid with a measured period. It is usually not very difficult to determine the periods of the light variations, and the absolute magnitudes can then be read off from this period–luminosity relation. The apparent visual magnitude can be measured and the distance modulus can be determined, which tells us the distance for instance to the Large Magellanic Cloud or to the Andromeda galaxy.

The calibration of the period–luminosity relation can be achieved by means of Cepheids which are members of open clusters in our Galaxy. The distances to such clusters can be obtained by means of spectroscopic parallaxes, or by main sequence fitting, as discussed in Section 5.4.

The period–luminosity relation for the Cepheids has been extremely important for the determination of the distances to the nearest extra-galactic systems in our local group of galaxies.

In Fig. 16.7 we show the period–luminosity relation for population I Cepheids, as determined by Sandage and Tammann (1969), Schmidt (1984), and by us (1986)*. There is still an uncertainty of about 0.5 magnitudes in the absolute calibration, at least for the longer period Cepheids, which are the most important ones, because they are the brightest which can be observed best in extragalactic star systems. Discussions about the accurate cluster distances are still going on. The problem is the uncertainty in the trigonometric parallaxes for the nearby stars upon which all our distance determinations are based. It is very difficult to measure high precision trigonometric parallaxes if the parallax angles are less than 0.05 arcsec!

How can we understand the existence of the period–luminosity relation?

* There is actually a $B - V$ dependent term in the period–luminosity relation. We have used the average relation between $(B - V)_0$ and the period to eliminate this term.

Intuitively, it seems plausible that it should take the more luminous stars, which are the largest stars, longer to expand and contract than it takes the less luminous, smaller ones. Actually, the periods of the pulsating stars are determined by a resonance phenomenon. We are essentially observing a standing wave. The phenomenon can be compared to a rope fastened at one end to a wall. If you move the free end of the rope, you can generate a limited number of standing waves with very distinct frequencies, the eigenfrequencies. The wave which you generate by moving the free end proceeds along the rope until it is reflected at the fixed end. We observe standing waves if at

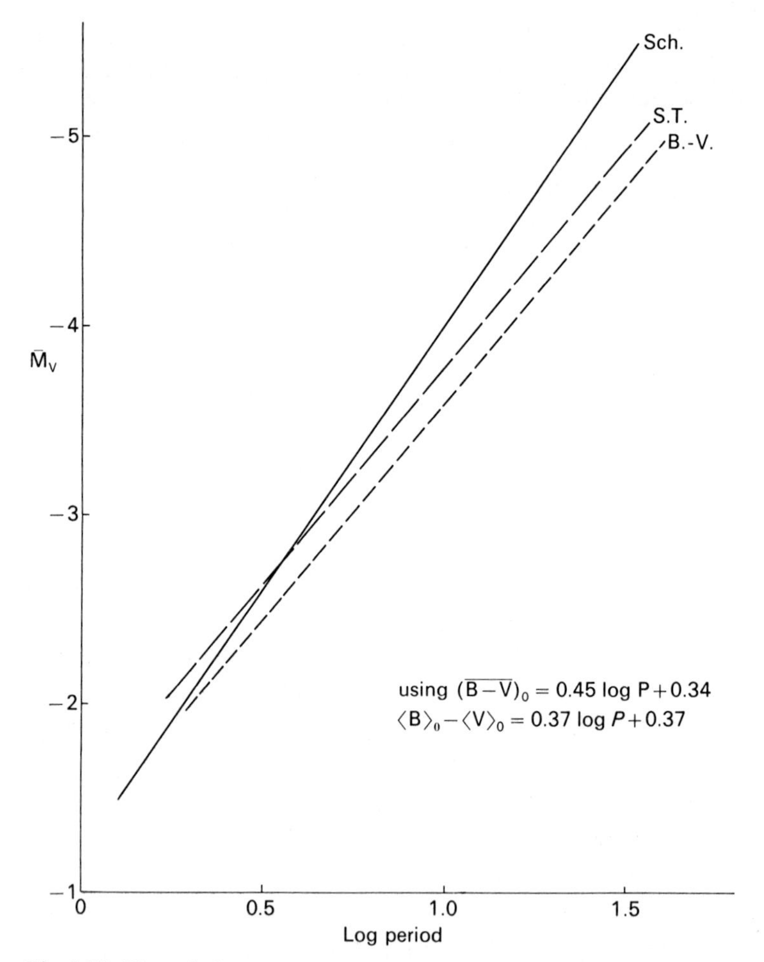

Fig. 16.7. The relation between absolute magnitudes and periods (period–luminosity relation) is shown for population I Cepheids according to different recent studies. S.T. stands for Sandage, Tammann 1969. B.-V. stands for Böhm-Vitense 1986, who adopts the Sandage, Tammann relation, except for a decrease in intrinsic brightness by 0.2 magnitudes. Sch. stands for Schmidt 1985, who redetermined the distances to clusters with Cepheids.

any given point the reflected wave has the same phase as the oncoming wave. We can observe a low frequency, the fundamental mode, and we can generate some higher frequency waves, the overtones. The star acts in very much the same way as this rope. The center of the star represents the fixed end of the rope because the high-density center cannot move. The surface of the star is the free end of the rope which, due to some excitation mechanism, is moving back and forth. Only if the reflected wave has the same phase as the inward moving wave, can the star generate a standing wave or a radial pulsation. Clearly, the travel time from the outside in and back is very important for the determination of these resonance waves. The travel time through the star depends on the density and temperature stratification in the star. This is why the frequencies of pulsation can tell us so much about the interior structure of the star.

In a qualitative way, the fundamental period P can be estimated from the condition that the period P is given by the travel time of a pressure wave to the centre of the star and back, i.e.,

$$P \sim \frac{2R}{C_S} \tag{16.4}$$

where C_S is the velocity of sound. With

$$C_S = \sqrt{\left(\gamma \cdot \frac{P_g}{\rho}\right)} \quad \text{where} \quad \gamma = \frac{C_p}{C_v} \sim 5/3 \tag{16.5}$$

for a monatomic gas. C_p and C_v are the specific heats at constant pressure and constant volume, respectively. At any place in the star the pressure must balance the weight of the overlying material. We derive for the center.

$$P_g \approx \bar{\rho} \cdot R \cdot \bar{g} \approx \bar{\rho} \cdot R \cdot \frac{GM}{R^2} = \bar{\rho} \cdot \frac{G \cdot M}{R} \tag{16.6}$$

where G is the gravitational constant and M the mass of the star. $\bar{\rho}$ is the average density and $\bar{\rho} \cdot R$ is the mass in a column of 1 cm^2 cross-section above the center. This means

$$\bar{P}_g / \bar{\rho} \sim \frac{GM}{R}.$$

We then find for the period

$$P \sim 2R \Big/ \sqrt{\left(\gamma \cdot \frac{GM}{R}\right)} \sim \frac{2}{\sqrt{\gamma \cdot G}} \cdot \sqrt{\left(\frac{R^3}{M}\right)} \tag{16.7}$$

or

$$P \sim \frac{2}{\sqrt{(\gamma \cdot G)}} \cdot \frac{1}{\sqrt{\rho}} \cdot \sqrt{\left(\frac{3}{4\pi}\right)} \tag{16.8}$$

or

$$P \sim \frac{2}{\sqrt{(\gamma \cdot 6)}} \cdot \sqrt{\left(\frac{3}{4\pi}\right)} \cdot \rho^{-1/2} \propto \rho^{-1/2} \tag{16.9}$$

or

$$P = \text{const. } \rho^{-1/2}. \tag{16.10}$$

This qualitative estimate gives only the correct dependence on density. The constant has to be determined by more accurate calculations. Equation (16.9) explains the observed period–luminosity relation. The larger the luminosity, i.e., the radius, the smaller the density, and the longer the period.

17

Explosive stars

17.1 Supernovae

17.1.1 Classification of novae and supernovae

The ancient astronomers had already noted that sometimes new stars became visible in the sky and after some time disappeared again. In the Middle Ages the astronomers called these stars novae, which is the Latin word for new stars. Some of these new stars were exceedingly bright, and were later called supernovae. Three of these supernovae were observed in historic times: Tycho de Brahe's supernova, which occurred in the year 1572, Kepler's supernova which became very bright in the year 1604, and a supernova which was observed by Chinese astronomers in the year 1054. At the location of the Chinese supernova we now see the Crab nebula in the constellation of Taurus. The nebula got its name from its appearance which reminds us of a crab. The Crab nebula still expands with velocities of about 1400 km s^{-1}, showing that a truly gigantic explosion must have occurred 900 years ago.

What are these novae and supernovae? How often do they occur? What kinds of objects are their progenitors? What leads to such gigantic explosions? What distinguishes novae and supernovae? Are all supernovae similar events or do we have to distinguish different kinds of novae or supernovae? These are questions for which we would like to find the answers.

Both novae and supernovae are objects which suddenly increase their light output by many orders of magnitude. They are usually not visible before the outburst, so they appear as new stars. Since we do not see them before the outburst, we really do not know directly what types of objects the progenitors are, except for the very unusual supernova which was discovered in the Large Magellanic Cloud in February 1987, for which the progenitor has been observed before and was classified as a B3 Ib supergiant. This supernova was very unusual, not only because it was intrinsically much fainter than other supernovae observed so far, but also because the time dependence of its light output was very different from that of other known supernovae. Other supernovae therefore probabhy have very different progenitors.

In Fig. 17.1 we reproduce a plot by W. Zwicky (1965), showing the frequency of new stars with different maximum brightnesses. We see a continuous distribution which is determined by both the intrinsic brightnesses of the novae and their distances. If we want to know the distribution of the intrinsic brightnesses we have to compare objects at nearly equal distances, such as, for instance, the novae or supernovae in the Andromeda galaxy or all the ones in the Magellanic Clouds. Supernovae are not very frequent events, so we cannot observe supernovae in only one galaxy, but we have to combine observations from many galaxies at approximately the same distances. In Fig. 17.2 the apparent magnitudes of the observed new stars in nearby extragalactic systems have been plotted. In this graph two groups can clearly be distinguished; the brighter group being the supernovae and the fainter one of the novae.* Novae are less bright and more frequent. They can best be observed in nearby galaxies and in our own Galaxy. The one supernova observed in the Andromeda nebula, also called M31, had a maximum apparent magnitude m_v between 7 and 8, while the supernova

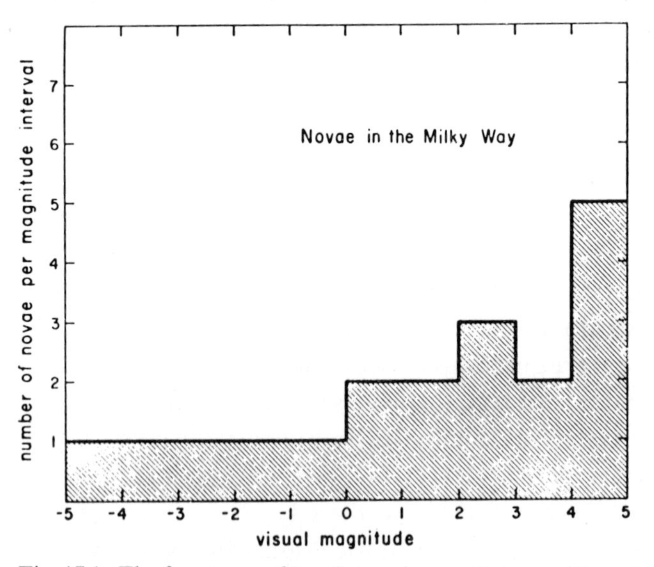

Fig. 17.1. The frequency of 'new' stars in our Galaxy with a given apparent magnitude is shown for the last 700 years, according to Zwicky (1965).

*On average the supernovae plotted here are further away then the novae shown, The difference in intrinsic brightness is therefore still larger than indicated by Fig. 17.2.

1987A observed in the Large Magellanic Cloud reached a maximum apparent brightness of 2.9 magnitudes. It clearly appeared much brighter than any other extragalactic supernova observed so far.

With a distance of $700\,\text{kpc} = 7 \times 10^5\,\text{pc}$ the absolute magnitude at maximum light for the Andromeda supernova was (neglecting interstellar absorption)

$$M_\text{v} = m_\text{v} - 5 \times 4.85 = m_\text{v} - 24.3 = 7.3 - 24.3 = -17.$$

For the Large Magellanic Cloud 1987A supernova we find with a distance modulus of $m_\text{v} - M_\text{v} = 18.3$ an absolute magnitude at maximum visual light of

$$M_\text{v} = 2.9 - 18.3 = -15.4$$

This has to be compared with the absolute magnitude $M_\text{v} = -8$ for the brightest 'normal' star in the galaxy, or with the solar absolute magnitude of $M_\text{v} = +4.8$. The supernova is 23 magnitudes brighter, or as bright as 10^9 suns, in the visual. Actually, the average M_v for supernovae appears to be $M_\text{v} \sim -18 \pm 1$.

According to the records of the Chinese astronomers, the Taurus supernova reached a maximum apparent visual magnitude of about -5, which means it was visible during daytime. For Tycho's supernova the maximum m_v was about -4 and for Kepler's about -3. These magnitudes

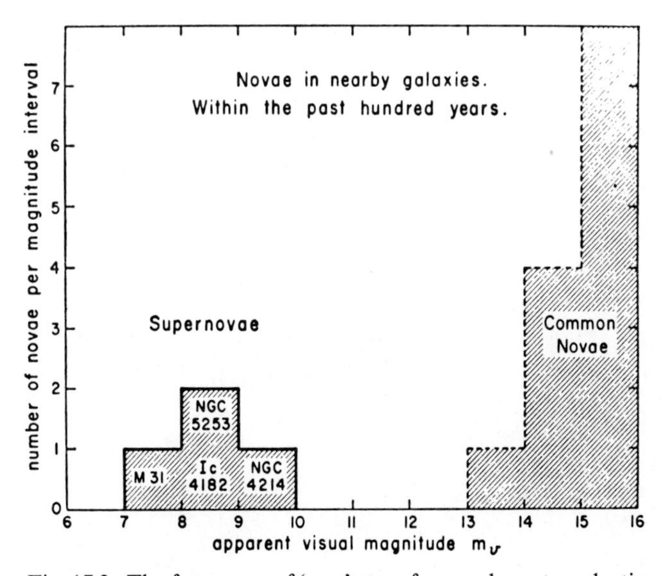

Fig. 17.2. The frequency of 'new' stars for nearby extragalactic systems with a given apparent visual magnitude is shown, according to Zwicky (1965).

are, of course, influenced by different distances and by absorption due to the interstellar gas and dust which is especially large for Kepler's supernova. The brightness of a supernova decreases exponentially with time. After one or two years it becomes invisible.

Altogether, several hundred supernovae have been observed. Unfortunately, none has occurred in our own Galaxy lately. The maximum brightnesses of the supernovae observed in external galaxies are about two magnitudes fainter than the whole galaxy. For the smaller, fainter galaxies the supernova can be just as bright as the whole galaxy. In Fig. 17.3 we show a photograph of a supernova in an external galaxy, IC 4281.

According to Zwicky, who has observed most of the supernovae, there are at least five different types as judged by their light curves. Most of the observed supernovae belong, however, to two types, called type I and type II. They can be distinguished by the shapes of their light curves as well as by their spectra. The observed light curves of some supernovae of type I are shown in Fig. 17.4. A typical type I supernova declines by 15–19 magnitudes during the first two years after the outburst, or by about 8 magnitudes in one year.

Figure 17.5 shows some light curves of type II supernovae. They generally look much more irregular than those of type I, although some do not look

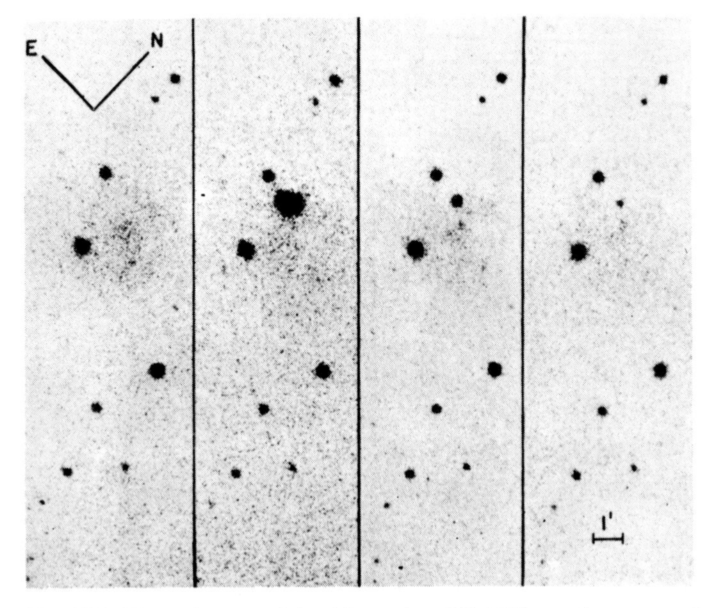

Fig. 17.3. Two pictures of the faint galaxy IC 4281 are shown, one before the outburst of the supernova and one after the occurrence of the supernova. (From Zwicky 1965.)

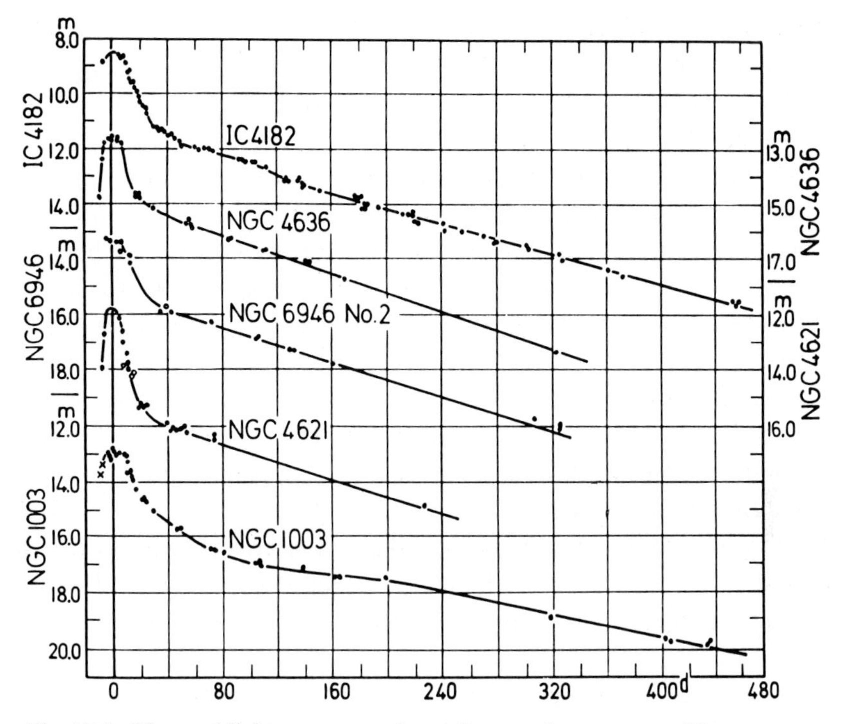

Fig. 17.4. Observed light curves are shown for type I supernovae. The names of supernovae are given. (From Unsöld 1982.)

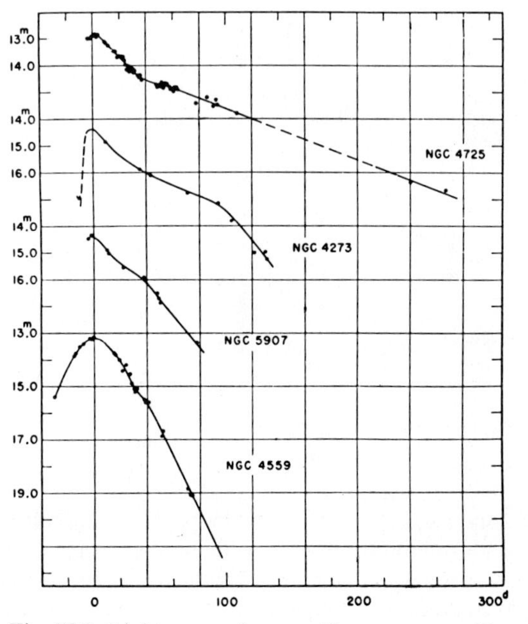

Fig. 17.5. Light curves for type II supernovae. (From Zwicky 1965.)

too different. When looking at such light curves, we have to be careful with the interpretation.

The spectra of the type I supernovae are still not understood. They show broad bands (see Fig. 17.6) and it is hard to decide whether the light regions are emission lines or whether the dark bands are absorption bands in a bright continuum. We probably have both absorption and emission lines. It seems that in type I supernovae we see broad absorption lines of heavy elements while in type II supernovae hydrogen absorption lines are identified. The new classification distinguishes between types Ia and Ib. In type Ib supernovae HeII lines are seen. If the widths of the lines are interpreted as being due to Doppler shifts, then they indicate velocities of up to 20 000 km s^{-1}, or $\sim$15 000 miles s^{-1}. The material would fly halfway around the whole Earth in one second. It appears that we are watching gigantic explosions.

A few days after maximum light the overall spectral energy distributions resemble that of a supergiant with temperatures around 10 000 K at maximum and cooling down to 5000 or 6000 K after two or three weeks.

In Fig. 17.7 we show the spectrum of a type II supernova. Broad lines of Hβ, Hγ, and Hδ are seen. The line widths are also large, corresponding to

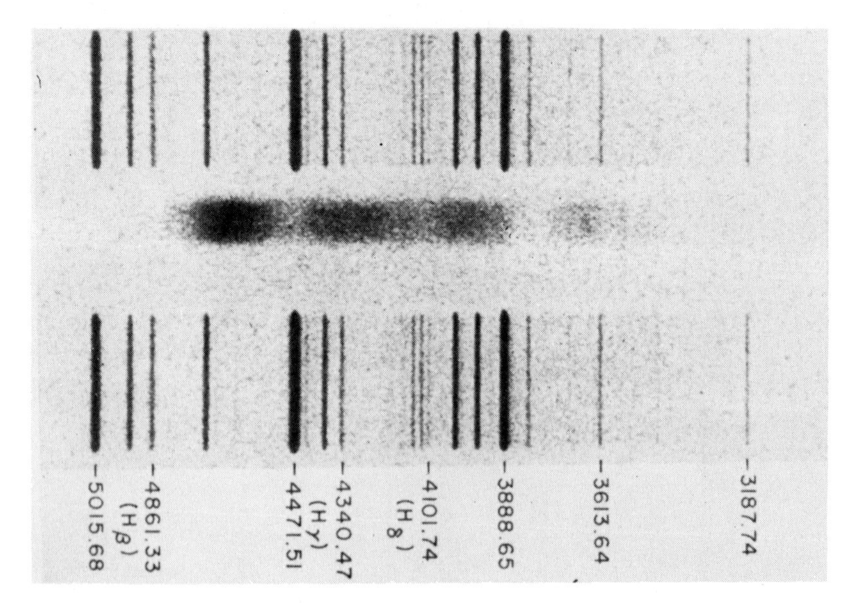

Fig. 17.6. We show a spectrum of a supernova of type I. It is difficult to decide whether we see broad absorption bands in a continuous spectrum, or whether we see broad emission bands on a much lower continuum. A comparison spectra with hydrogen and helium emission lines are also shown on both sides of the supernova spectrum. (From Zwicky 1965.)

velocities of up to about 10 000 km s^{-1}. The velocities seem to decrease as time goes on. Whether this means an actual slowing down of the motion or whether it means that we see successively deeper layers with decreasing velocities is not quite clear.

The Large Magellanic Cloud supernova also showed hydrogen and helium lines. The hydrogen lines were in the beginning shifted towards the blue, corresponding to Doppler shifts for 20 000 km s^{-1}, but then slowing down to velocities of about 10 000 km s^{-1} and remaining at the same velocity for several weeks.

In the first days this supernova had an overall energy distribution resembling that of a hot star but deep broad absorption troughs seemed to

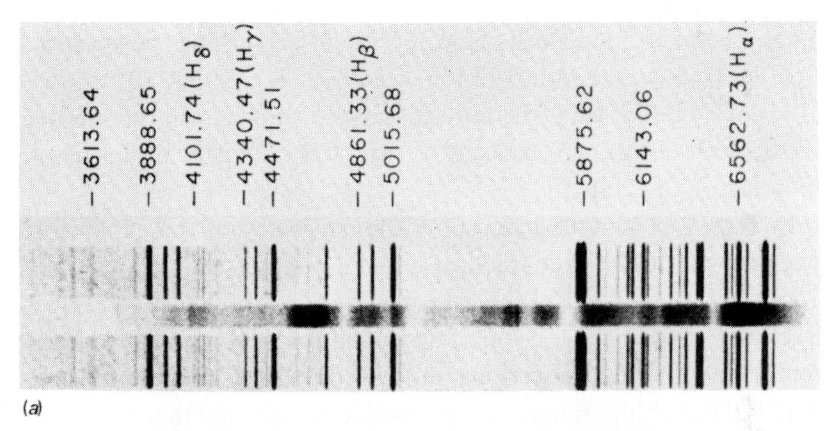

(*a*)

Fig. 17.7(*a*). The spectrum of a type II supernova is seen. The strong absorption lines can be identified with lines of hydrogen or helium, seen in the laboratory comparison spectra on the side of the stellar spectrum. (From Zwicky 1965.)

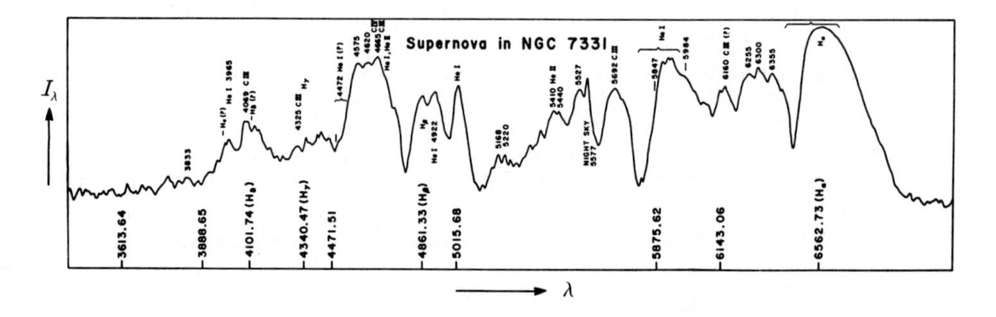

(*b*)

Fig. 17.7(*b*). The intensities I_λ measured in the spectrum of a type II supernova are shown as a function of wavelength λ. The broad Hα emission line at 6562.7 Å can well be recognized, as well as the emission for the He I line at 4471.5 Å.

be superimposed. After about two weeks the overall energy distribution suggested temperatures around 6000 K.

There seems to have been one other supernova which was also observed shortly before maximum light, when its overall energy distribution suggested a temperature around 20 000 K. At maximum light the temperature appeared to have dropped to 10 000 K, but no lines were seen.

It looks as if the supernovae become very hot briefly and then start expanding while they cool at the surface, probably due to the expansion.

After the maximum the supernovae appear to be rather cool, therefore recent studies have attempted to identify the broad, deep absorption bands with lines of Fe II, Ca II, Na I, Mg II, etc., all lines which are normally seen in rather low temperature stars with temperatures between 5000 and 10 000 K. There seems to have been some success in explaining the spectra.

After 10 days the radius of the ejected shell is about $10^9\ \mathrm{cm\,s^{-1}} \cdot 10^6\ \mathrm{s} = 10^{15}\ \mathrm{cm} = 60\ \mathrm{AU}$, which is more than the distance from the sun to Pluto. If the explosion started from a star of, say, $10\ R_\odot$, then the radius has increased by a factor of 1200.

In Figs 17.8 and 17.9 we show the geometry of the expansion. In the beginning a rather large fraction of the material (section *d*), in our drawing about 10%, is in front of the star and can absorb light. The corresponding fraction *c* in the back is covered up by the star and is invisible. The material on the sides of the star (sections *a* and *b*) contributes to an emission line. This material is moving transversely to the line of sight and therefore the emission

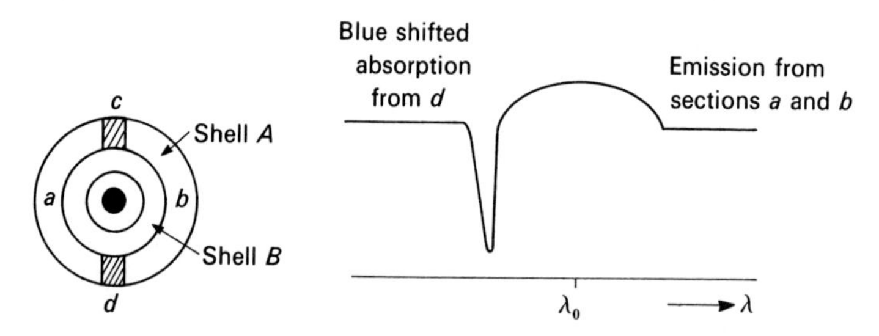

Fig. 17.8. In the beginning of the supernova expansion the material in the outer shell (shell *A*) is rather dense. There is much material in section *d* in front of the star to absorb light from the deeper layers, causing a blueshifted absorption line. Sections *a* and *b* cannot contribute to the absorption line; they can only emit light, giving rise to an emission line. These sections move on average transverse to the line of sight; the emission line is therefore only broadened but not shifted. Redshifted emission lines are emitted from section *c* but are obscured by the star. The expected absorption–emission line profile is shown on the right side of the figure. These types of line profiles were first observed for the star P Cygni (which is not a supernova) and are therefore called P Cygni lines. In Fig. 17.10 we show a part of the spectrum of this star.

lines have essentially no Doppler shift. The expected line profile is shown qualitatively in Fig. 17.8 on the right. Shell *B* below shell *A* contributes only to the background against which *A* is seen in absorption.

Sometime later we have the situation shown in Fig. 17.9. Shell *A* has expanded so far that only a small fraction *d* is in front of the star and can absorb light from the star. The density has decreased by a large fraction. Shell *A* contributes little to the absorption and emission. It is now transparent and shell *B* becomes visible. Shell *B* has a smaller velocity that *A*, so the absorption lines caused by shell *B* show a smaller Doppler shift and the spectrum shows a smaller shift in absorption and emission, although the expansion did not actually slow down. This only shows how careful one has to be when interpreting changes in the velocity shifts of emission and absorption lines.

In Fig. 17.10 we show the spectrum of the star P Cygni, which is not a supernova, but which also has an expanding shell, but the velocities are much smaller for this star. We see the typical combination of absorption and emission lines.

Fig. 17.11 demonstrates that we also have to be very careful when we try to interpret the change of light output by a supernova. The supernova may be

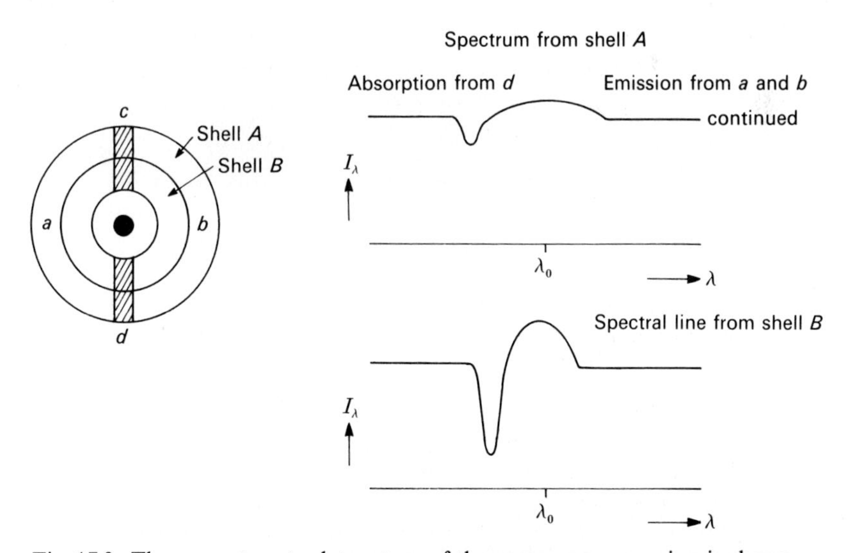

Fig. 17.9. The geometry at a later stage of the supernova expansion is shown. Shell *A* has greatly expanded and becomes transparent because of its low density. There is now much less material in section *d* in front of the star, which can cause a blueshifted absorption line. The next deeper shell *B* now gives the main contribution to the emission and absorption. Shell *B* has lower velocities than shell *A*. The expected line profiles due to shells *A* and *B* are shown on the right-hand side. The observed line profile is a superposition of these profiles.

in the neighborhood of dust clouds. When the supernova explodes light will be emitted in all directions. At first, we only see the light which is emitted into our direction (see Fig. 17.11). Later on we may also see light which is reflected from a neighboring cloud. We see then a so-called light echo.

17.1.2 Estimation of age and distance of supernova remnants and mass loss of supernovae

If the ejected mass expands so fast, we should be able to see the shell growing. Suppose the supernova is at a distance of 2000 pc, the probable distance of the Crab nebula. Then a shell of 1 AU diameter would have an angular diameter of 1/2000 arcsec. (At a distance of 1 pc, the diameter of 1 AU has an angular diameter of 1 arcsec, by the definition of the pc.) With an

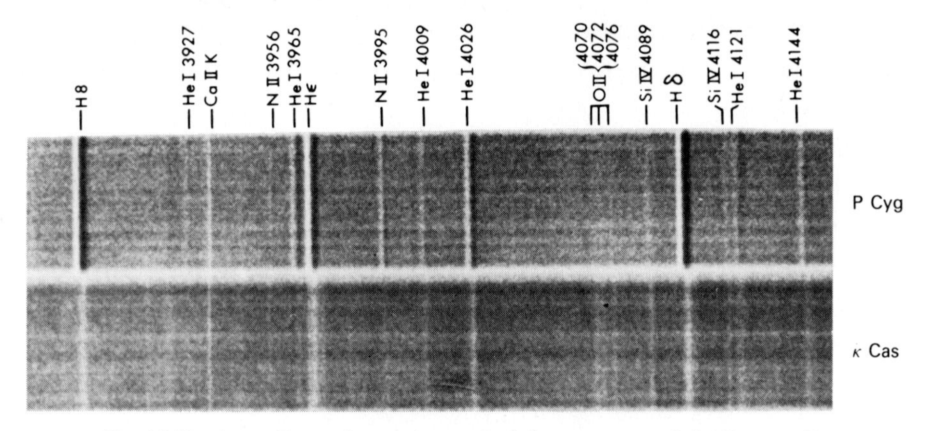

Fig. 17.10. A small wavelength interval of the spectrum of the Be star (Be means B star with emission lines) P Cygni is reproduced, showing the emission–absorption line profiles seen for an expanding gas shell around a star. For comparison the spectrum of the star κ Cas is also shown. (From Morgan, Abt & Tapscott 1978).

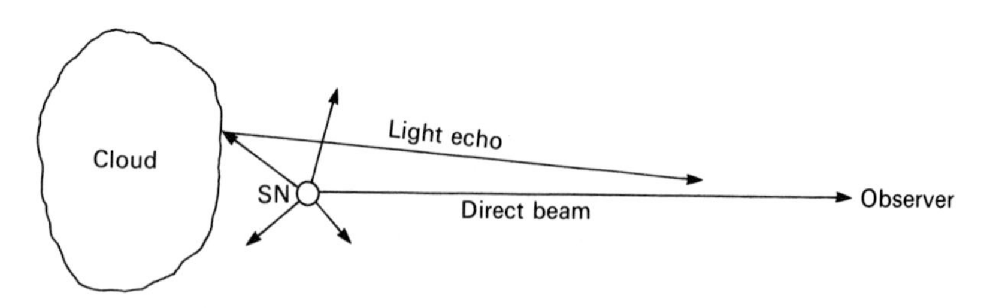

Fig. 17.11. The change of the light output of a supernova after maximum light is not only determined by the light output of the supernova itself, but may also be influenced by light reflected from nearby dust clouds, the so-called light echo.

expansion velocity of 10 000 km/s = 10^9 cm/s, as observed shortly after maximum, the shell would have a radius of 3×10^{16} cm, or 2000 AU after one year if the velocity remained constant. This corresponds to 1 arcsec if it is at a distance of 2000 pc. Actually, the expansion velocity of the Crab is observed to decrease. The present expansion velocity of the Crab nebula, for instance, is about 1400 km s^{-1}, or $\sim 5 \times 10^{15}$ cm per year. After 1000 years the nebula will then have a radius of 5×10^{18} cm, or 1.7 pc, or 3×10^5 AU. At a distance of 2000 pc, this corresponds to $\sim$ 150 arcsec. The actual radius is about 180 arcsec, showing that the velocity has decreased during the 950 years and was, on average, $\sim$ 2000 km s^{-1} if the Crab is at a distance of 2000 pc. The average velocity could hardly have been less than the present 1400 km s^{-1}, so the Crab nebula must be at a distance $d \geqslant 1400$ pc.

We can use this method in a more refined way to determine the distance to the Crab nebula and thereby the maximum absolute magnitude of the Crab supernova. The present expansion of the nebula is 0″.222 arcsec per year. The measured radial velocity, i.e., perpendicular to the direction in which we measure the expansion, is 1450 km s^{-1}. If the velocity in the two perpendicular directions is the same, then the Crab must be at a distance of 1380 pc. At this distance the supernova would not have been very bright intrinsically, only $M_V = -15.7$, if there were to be no interstellar absorption, or with 1.5 magnitude interstellar absorption, as seems reasonable (see Chapter 19) an absolute magnitude of $M_V = -17.2$ is obtained.

The expansion may, however, not be spherically symmetric. If the expansion velocity perpendicular to the line of sight is 1.25 times larger than the radial velocity in the line of sight, then the distance becomes 1730 pc. At such a distance it is reasonable to assume that the interstellar absorption dims the supernova by about 2 magnitudes. Taking this into account, we find a maximum absolute magnitude for the Crab supernova of $M_V = -18.2$, which would be close to the maximum brightness observed for other extragalactic supernovae of type I.

From the reconstructed light curves it seems very likely that the Chinese nova and Kepler's and Tycho's novae were all supernovae of type I. There seems to be a spread in maximum M_V for type I supernovae. The brightest ones are called type Ia. Since all Galactic ones appear to have been somewhat fainter they may have been of type Ib, or they may have been more obscured or further away than we think.

How much mass is ejected in such a supernova explosion? From the light emitted by the Crab nebula now, an estimate of the luminous mass can be made which gives a range $0.35\, M_\odot < M < 3.5\, M_\odot$, which really does not tell us too much, except that it is not a negligible fraction of the mass of the supernova. Part of the luminous mass in the nebula was swept up from the interstellar medium. If the density was 1 hydrogen atom per cm^3 then the accreted mass was $\sim 0.2 M_\odot$. Most of the mass is probably from the supernova outburst.

17.1.3 The origin of supernovae

Unfortunately nobody has ever observed a presupernova until recently when the supernova in the Large Magellanic Cloud exploded. In this case the star at the position of the supernova was previously observed and has been classified by Sanduliac as a B3 Ib supergiant. This supernova increased its brightness by about 10 magnitudes. For other supernovae we can only make rough estimates. Since the brightest normal stars have $M_V = -8$, a supernova must increase its brightness by at least 10 magnitudes, probably much more. If the Crab presupernova had $M_V = -8$ and at maximum it had $m_V = -5$, then the presupernova would have had $m_V = -5 + 11 = +6$. A trained astronomer would just barely have been able to see it with the naked eye, but it might have escaped observation. It was not recorded, probably it was fainter and therefore not visible. The other galactic supernovae were further away and more obscured by interstellar dust. Galileo had a telescope but he obviously had not seen the presupernova before the sudden brightening. Whether he could have seen it, we do not know. He did observe Kepler's supernova after it had become bright.

Three supernovae have been observed in nearby galaxies. From a comparison with the brightest stars in those galaxies, an absolute visual magnitude of $M_v = -18 \pm 1$ was derived.

We can get some information about the mass of the presupernovae if we check with which kinds of stars they are associated. We saw earlier that old stars are seen in globular clusters and they all have masses less than 1 solar mass. Elliptical galaxies supposedly have mainly such old stars. On the other hand, young massive stars are found mainly in the spiral arms of galaxies, of course together with less massive stars. If presupernovae are massive stars, we might expect to see them associated with young star clusters. Zwicky at the California Institute of Technology was the first one to start a systematic search for supernovae. He did some statistics with 111 supernovae, observed until 1965. He found that about 26% of all types I supernova occurred in elliptical galaxies, i.e., among old stars. Other type I supernovae were found in spirals and irregular galaxies in about the proportions of the number of existing galaxies. In the spirals, the supernovae are found mainly in the outer parts. Since a rather large number of type I supernovae occur in elliptical galaxies, those must belong to the old population. The progenitors are probably low mass stars.

Type II supernovae have not been seen in elliptical galaxies, it seems, but even that does not seem to be quite established. The classification is sometimes difficult; a few supernovae seen in ellipticals could not be classified very well because not enough observations existed. So it is not quite clear at the moment, but it appears that type II supernovae are not found associated with ellipticals, suggesting that they belong to a young stellar population and probably have massive stars as progenitors. In external spiral galaxies type II supernovae often appear to be associated with regions containing the massive O and B stars. If they indeed have 5-20 $M_\odot$ stars as progenitors, (the LMC supernova progenitor appears to have had 15 $M_\odot$) there should have been several supernovae explosions in the Pleiades and in the Hyades star clusters. Should we be able to see the remnants? How old are the oldest remnants that we actually see? Some spherical nebulae are supposed to be supernova remnants like the Cygnus loop, the Veil nebula, the Vela remnant, or the Gum nebula (see Figs 17.12 and 17.13).

Fig. 17.12. The Crab nebula (in the constellation of Taurus) is shown. Near the center of the nebula there are two stars close together. The lower right star of these two shows a featureless blue continuous spectrum. It is the pulsar. (From Abell 1982.)

If type II supernovae have massive stars as progenitors, should we not expect to see supernovae or at least supernovae remnants in young galactic clusters with many massive stars?

From the observed radius and the radial velocities, the ages of the remnants can be estimated, as discussed above. They are of the order of 10^4–10^5 years, and none is older than 10^5 years. If there were perhaps five massive stars in the Pleiades which became supernovae during the lifetime of 10^8 years, there was one every 10^7 years. The chance of seeing a remnant now are thus 1 in 100, which is consistent with the fact that we do not see any. There are not enough young clusters around for us to have a good chance of seeing one in a cluster. So that does not tell us much.

There are, however, young, very populous clusters in the LMC. Perhaps we have a chance to see a supernova go off in one of them. It would be very interesting to study the effects on the other cluster stars. The observed LMC supernova was not in a cluster.

1.7.1.4 Energetics of supernova explosions

How much energy is involved in a supernova explosion? How can this energy be supplied?

The most exciting thing about the supernovae is the enormous energy

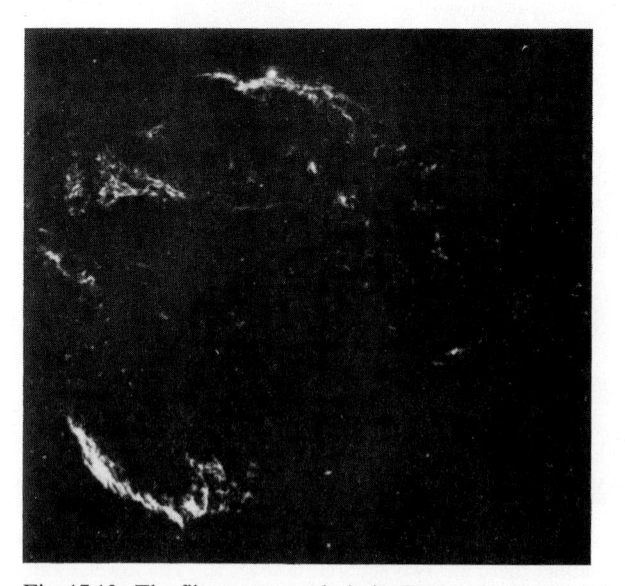

Fig. 17.13. The filamentary nebula in Cygnus is shown. It is believed to be the remnant of an old supernova. (From Abell 1982.)

output. If its absolute visual magnitude is $M_V \sim -19$, as compared to the solar absolute visual magnitude of $M_V \sim +5$, we see that the visual light output of the supernova is about $10^{9.6}$ times as much as that of the sun. Considering that the supernova has $T \geqslant 10\,000$ K and has a bolometric correction say of about 0.4 magnitudes, we find that the supernova at maximum has a luminosity which corresponds to at least 10^{10} suns.

This is about a factor of 10 less than the luminosity of the whole galaxy. We saw above that a supernova at maximum is only two magnitudes fainter than a bright galaxy.

The luminosity at maximum is then 10^{43}–10^{44} erg s^{-1}, which means the total visible energy output is 10^{49}–10^{50} erg. It is generally assumed that the energy output in the form of neutrinos is even higher, and the neutrinos observed after the outburst of the LMC supernova seem to confirm this.

If a star were to convert all its hydrogen into helium by nuclear fusion, then 0.7% of its mass would be converted into energy and the total amount of nuclear energy liberated would be (see Volume 3)

$$E_{\text{nuclear}} = 7 \times 10^{-3} : c^2 \cdot M,$$

where M is the hydrogen mass of the star. For the sun this would be

$$E_{\text{nuclear}_\odot} = 1.3 \times 10^{52} \text{ erg}.$$

In an outburst, the supernova releases in 10 days almost as much energy as the whole nuclear energy available to a star with 1 $M_\odot$. But it is our present understanding that the supernova explosion happens at the end of the stellar evolution and therefore most of the nuclear energy has been used up already. There must be another energy source.

The gravitational energy released by a star when it collapses from a dust cloud into a star with radius R is

$$E_g \approx \frac{GM^2}{R},$$

which for the sun comes out to be $\sim 10^{48}$ erg, which again is not enough for a supernova.

If, however, the sun were to collapse to a white dwarf with a radius 100 times smaller than that of the sun, then the gravitational energy release would be 10^{50} erg. We would be already coming close to the energy release of a supernova. If the sun were to collapse to a neutron star with $R \leqslant 70$ km $= 7 \times 10^6$ cm, the total gravitational energy release would be $\geqslant 10^{52}$ erg. The gravitational collapse of a normal star into a neutron star could provide enough and even more energy than observed for a supernova explosion. In fact this seems to be the only known energy source which can provide enough energy.

17.1.5 The Crab pulsar

Does a presupernova collapse into a neutron star?

The only way to answer this is to look at supernova remnants like the Crab. In the center of the Crab nebula we see two stars (see Fig. 17.12). The one star is a normal F star. The other one emits a blue continuum with no spectral features. It has long been suspected to be the remnant of the supernova outburst. The real confirmation came only when it was discovered that it is a pulsar. This means it was observed to have pulsed light emission, with a period of 1/30 of a second. These pulses were first detected at radio wavelengths, but they were later also confirmed in the visual.

In Fig. 17.14 we show a series of pictures of the Crab nebula taken at different phases of the pulse cycle. The first picture was apparently taken shortly after the off phase of the pulses, the third and fourth and the ninth and tenth ones were taken during the on phases. The star with the F star spectrum is always visible, but the blue star with the featureless spectrum comes and goes, showing that the light is turned on and off 30 times a second. Because the period for these pulses is shorter than the resolution time of our eyes, which is 1/16 of a second, these light flashes remained undetected until the radio astronomers developed receivers with very high time resolution which enabled them to detect such short pulses of radio waves.

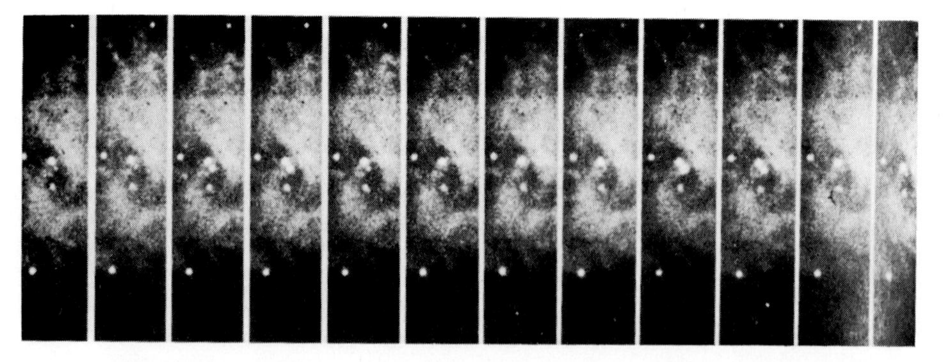

Fig. 17.14. A time sequence of photographs of the central pair of stars in the Crab nebula is shown as a function of phase in the pulse cycle. The time differences are less than 1/100 of a second. (Repeated exposures is successive cycles are necessary to get enough light.) While the F star remains constant in light, the pulsar turns on and off. (From Abell 1982. Photographs from Kitt Peak National Observatory.)

Between the main pulses there is always a smaller interpulse. Each pulse is subdivided into many small pulses with very short timescales. The main pulses occur extremely regularly. The period is very well defined, with no irregularities. The only phenomenon known to be so regular is rotation. But this means that this star must rotate 30 times a second. For such rapid rotation we must ask ourselves, how can the star stay together and not be torn apart by centrifugal forces? If the star does not disintegrate, the gravitational force has to overcome the centrifugal forces at the surface of the star, i.e.,

$$\frac{v^2}{R} \leqslant \frac{GM}{R^2} \quad \text{or} \quad R < \frac{GM}{v^2} \quad \text{where } v = \frac{2\pi R}{P} \tag{17.1}$$

P is the rotation period and v the equatorial rotational velocity. Inserting the expression for v, we find

$$R^3 < \frac{GM \cdot P^2}{4\pi^2} \approx \frac{7 \times 10^{-8} \cdot 2 \times 10^{33}\,\text{cm}}{40(30)^2} \sim 3 \times 10^{21}\,\text{cm}^3 \tag{17.2}$$

or $R \leqslant 1.5 \times 10^7$ for a $1\,M_\odot$ star or, in other words, the radius of the star must be less than 150 km.

The shortest possible period is obtained if we insert the highest possible velocity, namely $v = c$ where c = velocity of light, into the stability criterion (17.1). We then find

$$R < \frac{GM}{c^2} = \frac{7 \times 10^{-8} \cdot 2 \times 10^{33}}{9 \times 10^{20}} = 1.5 \times 10^5\,\text{cm} \tag{17.3}$$

and

$$P = \frac{2\pi R}{c} = \frac{10^6}{3 \times 10^{10}} \sim 0.3 \times 10^{-4}\,\text{s}. \tag{17.4}$$

In principle, much shorter periods than observed for the Crab pulsar would be possible. However, such objects would have extremely high densities. If we assume the Crab pulsar would have a radius of $\sim 100\,\text{km} = 10^7\,\text{cm}$, then the density for a $1\,M_\odot$ would be

$$\rho = \frac{2 \times 10^{33}}{4 \times 10^{21}} = 1/2 \times 10^{12}\,\text{g}\,\text{cm}^{-3}$$

If you pack nuclear matter densely, you find densities around $10^{14}\,\text{g}\,\text{cm}^{-3}$. Remember that for white dwarfs we found densities around 1 ton per cm^3. For the Crab pulsar we find, if its mass is about 1 solar mass, that the density is at least 1 million tons per cm^3. There is only one kind of star known, which can have such high densities, namely the so-called neutron stars. Because of the extremely high densities in these stars the electrons and protons are squeezed together to form neutrons. The stars therefore consist mainly of neutrons and have a radius of about 15 km.

How does the pulsar get such high rotational velocity?

It turns out that this is actually no problem. If a star collapses, conserving angular momentum for each unit volume, i.e.,

$$\rho \bar{r}^2 \cdot \omega = \text{const}, \tag{17.5}$$

then

$$\omega \propto \frac{1}{r^2}.$$

Shrinking $\bar{r}$ by a factor

$$\frac{1\,000\,000\,\text{km}}{100\,\text{km}} = \frac{1.5\,R_\odot}{100\,\text{km}} = 10^4$$

increases ω by a factor of 10^8 or the period $P = 1/\omega$ would decrease by a factor of 10^8. If the star originally had a rotational period of $1/30 \times 10^8\,\text{s} = 3 \times 10^6\,\text{s}$ or 1 month like the sun, then there is no problem.

Consider, for instance, a B star with $v_{\text{rot}} \sim 200\,\text{km s}^{-1}$, which corresponds to a rotational period of

$$P \sim \frac{2 \times 10^6\,\text{km} \cdot 2\pi}{v_{\text{rot}}} = \frac{1.26 \times 10^7\,\text{km}}{200\,\text{km s}^{-1}} = 6 \times 10^5\,\text{s} \tag{17.6}$$

or $P \sim 6$ days, or even a star with $v_{\text{rot}} = 20\,\text{km s}^{-1}$, which would have a rotation period of 60 days, they could easily have a rotational period as short as the Crab pulsar if they were to collapse into a neutron star.

For the Crab supernova, it then seems very likely that it gained its energy from the gravitational energy released in the collapse of a major fraction of its mass into a neutron star now visible as the Crab pulsar.

17.1.6 Other supernova remnants

The question then remains, do all supernovae explosions leave a neutron star? In order to answer this, we have to know how a neutron star can be discovered. In the Vela remnant, another pulsar is found. No pulsar has been seen in the Tycho and Kepler remnants, but this does not necessarily mean that there is no neutron star left.

The Crab pulsar emits light for only $\sim 10\%$ of its period. The only way to explain this is by assuming that the pulsar emits strongly beamed light (see Fig. 17.15). If the beam is narrow, the observer will be 'hit' by the beam only in exceptional situations. For the Crab, we actually see a pulse and an interpulse. We must be close to situation of Fig. 17.15(*b*). The Crab pulsar is the only pulsar seen to pulse in visual light. The radio beam is probably wider, so it can be more easily observed.

For the remnants in which we do not see a pulsar, the geometry may be like case (*c*) where we cannot see the beam. There may be neutron stars in all supernova remnants: we have no way of knowing. We are just lucky that the Crab pulsar beam points at us.

It is also possible that only certain types of supernovae leave behind a neutron star. Not knowing what the masses of presupernovae are, we must ask for which kind of star we might expect such an implosion–explosion process: which means implosion in the center leading to the formation of the neutron star, and explosion in the outer layers which leads to the explosion and the strong light increase.

In Volume 3 we will see that in the late stages of stellar evolution we may actually expect a situation when the interior of a massive star will become unstable leading to a collapse which will free enough gravitational energy to cause an explosion of the outer layers. We believe that such stars are the origin of type II supernovae.

17.2 Novae

17.2.1 Observations of novae

We shall now talk about the group of novae which is the less luminous but the more frequent. These are the so-called classical novae. They are much more frequent and have therefore been studied in much more detail. They reach maximum absolute visual magnitudes up to −9,

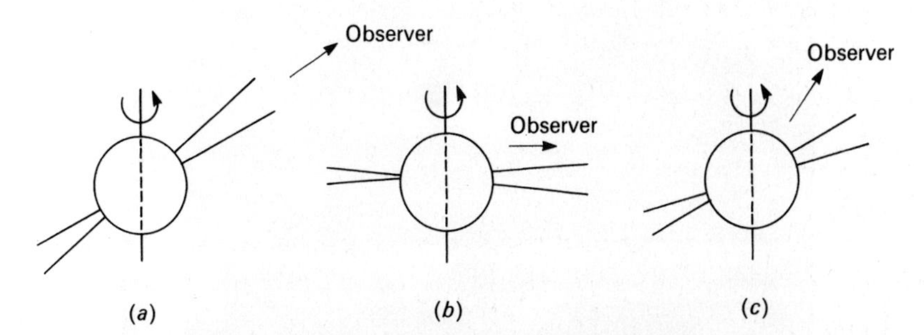

Fig. 17.15. If the pulsar emits strongly beamed light the observer may either see one light pulse per rotation period, case (*a*) or two pulses per rotation period, case (*b*) or she/he may not see a pulse at all, case (*c*), depending on the orientation of the light beam with respect to the rotation axis and on the direction of the line of sight.

i.e., about 10 magnitudes less than supernovae. Several of the novae have been observed before they erupted. Some of them have recurrent eruptions. During eruption they change their brightness by more than 9 magnitudes in a few days, they then decline to pre-eruption magnitude, sometimes within one or two years, like the supernovae, though the light curves and decline times are very different for different novae. There are fast, slow and moderate novae. They show a pre-maximum spectrum like O, B or A stars, but are definitely not main sequence stars, and are very subluminous. Their distances can again be determined from the radial velocities and the expansion of the nebula resulting from the explosion (see Fig. 17.16) in the same way as we discussed for the Crab nebula.

With this distance, we can determine the absolute magnitude of the prenova and see that it was a subluminous star. It is not faint enough for the light to be only due to a white dwarf. There is an additional light source.

Two observations were very important for the understanding of the nova phenomenon: (a) the nova DQ Herculis was found to be an eclipsing binary; subsequently more postnovae were found to be binaries, (b) several novae show small amplitude light variations with periods around 70 s.

Observations (a) suggest that all novae are close binaries. Observations (b)

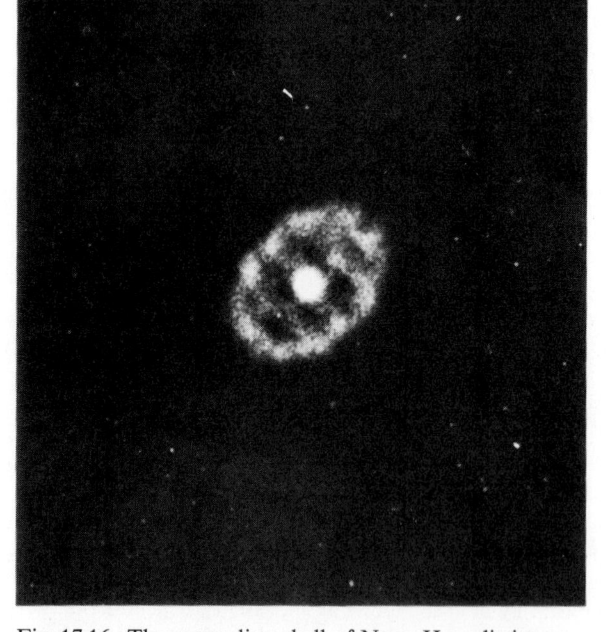

Fig. 17.16. The expanding shell of Nova Herculis is seen on this picture taken in 1972. The nova exploded in 1934. (From Abell 1982.)

indicate that one of the components must be a white dwarf, because only a white dwarf is small and dense enough to have oscillation periods of minutes. Normal stars would have longer periods, as was discussed earlier.

Since the post- and presupernovae are blue objects, brighter than a white dwarf, there must be an additional subluminous blue light source, which we think is a disk of hot material. In addition, there must be the second star, which has been seen in only one case, where it is a K2 IV star which means slightly more luminous than a main sequence star. In all other cases it is obviously too faint to be seen. It probably is a lower main sequence star.

17.2.2 Close binaries

Before we can go any further with our discussion of the origin of the nova explosions, we have to talk a little about close binaries. As we know, binaries exert gravitational forces on each other. If we replace both stars in our minds by point sources, then gravitational force and centrifugal force balance if they are in Kepler orbits, as we have discussed previously. But this is not the whole story, as we can see in Fig. 17.17.

Real stars are not point masses but extended objects. On star 2, with mass M_2 for instance, point A is closer to star 1 with mass M_1 than point O or point B. While at point O gravitational and centrifugal forces balance, this is not true of point A. At point A the gravitational force F_g working on 1 g is

$$\frac{G \cdot M_1}{(d_1 + d_2 - R_2)^2} = F_g \tag{17.7}$$

while the centrifugal force is the same for all points on M_1. (During one orbit, points O, B and A all describe a circle with the same radius d_2 if M_2 is not rotating.)

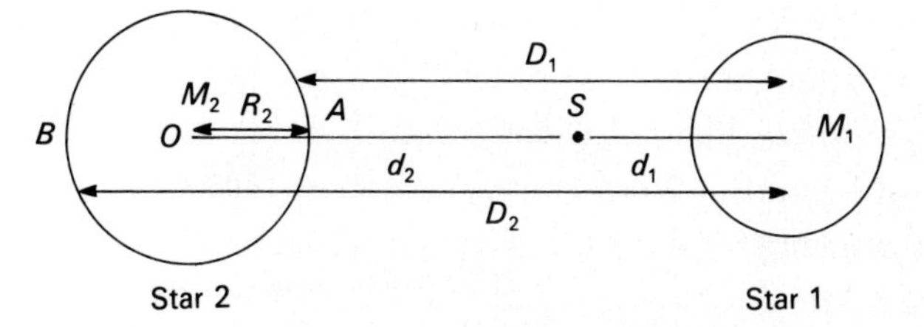

Fig. 17.17. For two stars in Kepler orbits gravitational and centrifugal forces balance for point masses at distance $d = d_1 + d_2$. For the matter at point A which is closer to star 1 the gravitational force is, however, larger than the centrifugal force; at point B the centrifugal force is larger than the gravitational force exerted by M_1. Point A is pulled toward M_1 while point B is pushed away.

For a Kepler orbit, the centrifugal and gravitational forces balance at point O, which means that if we write the equation for the forces working on 1 g of material we find

$$\omega^2 d_2 = \frac{M_1 \cdot G}{(d_1 + d_2)^2}, \tag{17.8}$$

we can then calculate, for point A, the difference between gravitational and centrifugal force $\Delta F(A)$, namely

$$\frac{GM_1}{(d_1 + d_2 - R_2)^2} - \omega^2 d_2 = \Delta F(A) \tag{17.9}$$

or

$$\Delta F(A) = \frac{GM_1}{(d_1 + d_2 - R_2)^2} - \frac{M_1 G}{(d_1 + d_2)^2} \tag{17.10}$$

according to (17.8). For $R_2 \ll (d_1 + d_2)$ we find from the Taylor expansion with

$$\Delta\left(\frac{1}{d^2}\right) = -2 \cdot \frac{1}{d^3} \cdot \Delta d$$

and

$$\frac{1}{(x - \Delta x)^2} = \frac{1}{x^2} + \frac{1}{\mathrm{d}x}\left(\frac{1}{x^2}\right)(-\Delta x) \tag{17.11}$$

that

$$\frac{GM_1}{(d_1 + d_2 - R_2)^2} \approx +2 \frac{GM_1}{(d_1 + d_2)^3} \cdot R_2 + \frac{GM_1}{(d_1 + d_2)^2} \tag{17.12}$$

and the difference in (17.11) becomes

$$\Delta F(A) = 2 \cdot \frac{GM_1}{(d_1 + d_2)^3} \cdot R_2. \tag{17.13}$$

This means we have a surplus gravitational pull on point A. At point B the situation is similar, though with the opposite sign. The gravitational force is now smaller by the same amount. At point B the gravitational force is less than the centrifugal force. The net effect is a surplus gravitational force at point A, pulling out point A from star 2, and a surplus centrifugal force pulling out point B from star 2. In other words, the gravitational pull with which point A and point B are attracted by star 2 is reduced, which makes the star more extended in the direction connecting the two stars. This is what we call the tidal bulges (see Fig. 17.18).

If the stars rotate once during one orbital period (as our moon does), then the tidal bulges always stay at the same place on the stars. However, if the stars have different rotational and orbital periods, then the bulges have to move around the stars during rotational and orbital periods. This takes some energy which has to come from the rotational energy. In fact, the bulges do not move quite fast enough; they lag behind because of friction. The deformed star then exhibits a torque on the other star in the sense that it tries to equalize the rotational and orbital periods. In the long run, rotational and orbital periods will become the same, but it takes time.

It can also be shown that the tidal forces will generally make the orbits circular. For close binaries which have been around for some time, we can therefore expect circular orbits and synchronized orbital and rotational periods. Both stars will then orbit their center of gravity with constant angular velocity ω.

Let us now ask the question, what will happen to a small third body, i.e., a dust grain or one asteroid or an atom P, which is also in the system.

Let us write down the equation for the forces acting on the mass point P. We describe the force $\mathbf{F}$ by means of the potential by setting

$$\mathbf{F} = \text{grad}\,\psi \tag{17.14}$$

which defines ψ. The surfaces for which $\psi = \text{constant}$ are called the equipotential surfaces. Equation (17.14) shows that along these surfaces there is no force component, the forces are always perpendicular to the equipotential surfaces. For the point P the gravitational potential with respect to each mass M is $\psi_g = GM/r$, where r is the distance to the mass M. The sum of the gravitational potentials for the point P is then

$$\psi_g = \frac{GM_1}{r_1} + \frac{GM_2}{r_2} \tag{17.15}$$

where r_1 and r_2 are the distances of P from M_1 and M_2 respectively (see Fig. 17.19).

In a rotating system we also have to consider the centrifugal force. Its potential is

$$\psi_c = +\frac{\omega^2}{2}\cdot r^2, \tag{17.16}$$

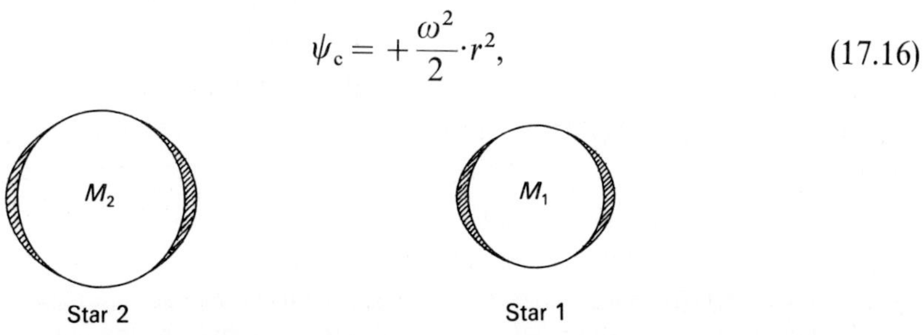

Fig. 17.18. Due to the geometrical extent of the stars there is an imbalance of gravitational and centrifugal forces at the 'inside' and at the 'outside' surfaces, leading to tidal bulges on the stars at the places where the surface material of the stars is closest and furthest apart.

where r is the distance from the center of gravity S, around which the whole system is orbiting. The total potential is therefore

$$\psi = \psi_g + \psi_c = \frac{GM_1}{r_1} + \frac{GM_2}{r_2} + \frac{\omega^2}{2} \cdot r^2. \tag{17.17}$$

The condition ψ = constant describes the equipotential surfaces. There are no force components in the equipotential surfaces. In the case of one spherical star when we consider only gravitational forces the equipotential surfaces are spheres around the center of the star. For a plane parallel atmosphere the horizontal planes are equipotential surfaces. There are no forces in the horizontal direction. The equipotential surfaces tell us in which direction the force acts. In the case of the two masses M_1 and M_2 working on the mass point P and in the presence of the centrifugal force the equipotential surfaces are rather complicated. In the neighborhood of each star they are almost spheres around each star, because the other star is far away and of little influence. The further away we go from each star the more distorted are the equipotential surfaces, as can be seen in Fig. 17.20. For one equipotential surface we find a crossing point between the two masses. This singular point is called the Lagrange point L_1. In the coordinate system of the orbiting binaries the forces on a point P in this location are zero, which means a mass point could reside in this location in theory for ever. If, however, a small disturbance brings it closer to the mass M_1 it will fall toward M_1, if a small disturbance brings it closer two M_2 it will fall towards M_2. A mass point at L_1 would be in a very unstable equilibrium. The equipotential surface which goes through L_1 is called the Roche lobe. We show in Figure 17.20 a plot of

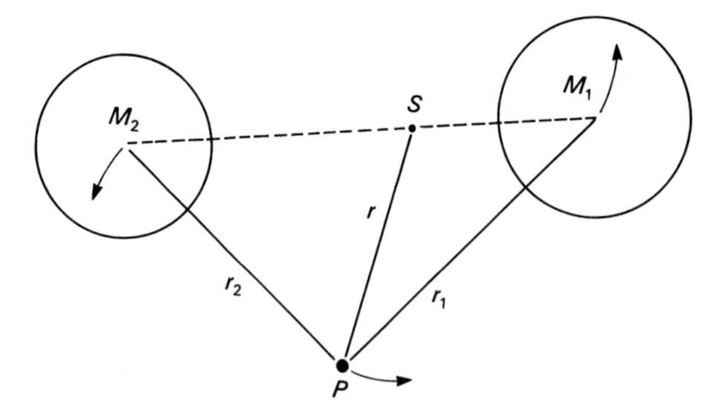

Fig. 17.19. In a close binary system with circular orbits around the center of gravity S, the gravitational potential at point P is given by the sum of the gravitational potentials for each star determined by the masses of the stars and the distances r_1 and r_2 from these stars. The potential for the centrifugal forces is determined by the distance r from the center of gravity around which the whole system is rotating.

the Roche lobe for the case that the mass ratio of M_1 to M_2 is 2 to 3.

This Roche lobe is very important for all discussions of close binaries. In the following we shall call the distance of M_2 from L_1 the length l_2 and the distance of M_1 to L_1 the length l_1. As long as the radius of M_2 is less than l_2 and the one of M_1 is less than l_1 nothing will happen, the stars are well separated. But suppose that M_2 is a white dwarf with $M \sim 0.8\, M_\odot$, and M_1 is an M star with $0.5\, M_\odot$. Then M_2 is well within its Roche lobe, but M_1 is much larger and may actually have a radius which is larger than l_1. In this case the star reaches beyond L_1 and the outer layer of the star is pulled over by M_2. The star loses mass to M_2. Such a system is called a semi-detached system. The material will not generally fall straight into the star M_2 but has some angular momentum and therefore spirals around the smaller star and forms a disk. We call this an accretion disk.

In the disk the matter does not flow in a strictly orderly way. The incoming material creates turbulence and friction. So some of the energy is transformed into heat, especially at the point where the incoming stream hits the disk. At that place, the so-called hot spot is formed, (see for instance Shu 1982, p. 194). But also within the disk there is some friction and the angular

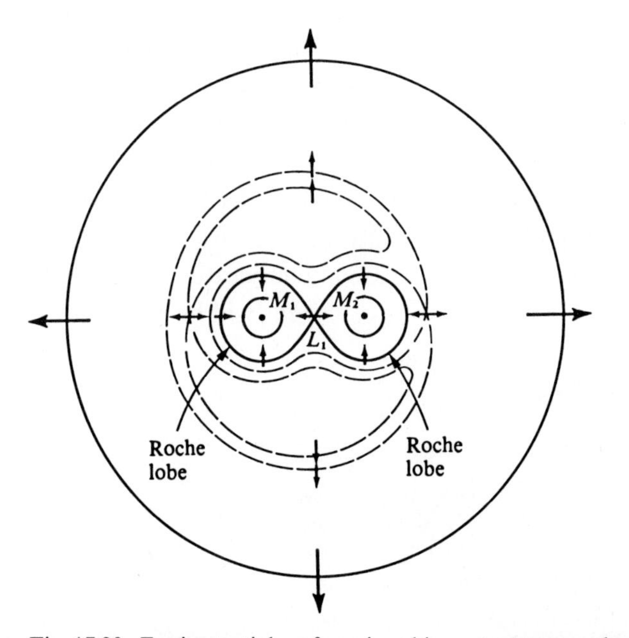

Fig. 17.20. Equipotential surfaces in a binary system are shown. The thick line shows the Roche lobe with the Lagrange point L_1. Near the stars the equipotential surfaces are nearly spheres. Outside of the Roche lobe they are rather complicated. The arrows show the direction of net force working on a particle at that location. At the Lagrange point L_1 the direction changes abruptly. (Adapted from Shu 1982.)

momentum will slowly be destroyed. The matter then falls into the white dwarf, as shown in Fig. 17.21.

This is the picture which people currently have about a prenova. What is seen is the light of the blue white dwarf plus the light of the hot disk and hot spot, making the object rather too bright for a white dwarf.

Suppose now that M_1 is a white dwarf and it collects matter coming from M_2, which is mainly hydrogen. The white dwarf itself consists mainly of carbon with a thin helium layer on the outside. (We will see in Volume 3 why this is so.) No nuclear burning takes place in the white dwarf. It actually cools down while losing energy at the surface. The white dwarf M_1 now accretes a layer of hydrogen from M_2 on its surface. The thicker this layer becomes, the higher will be the temperature at the bottom of this hydrogen layer. (For white dwarfs the temperature in the interior is still very high, possibly around 10^7 K). We remember that in a white dwarf the density is extremely high. Once the hydrogen atoms are no longer directly at the surface of the white dwarf they get into very high density regions. At such high densities the nuclear reactions start at lower temperatures than they would at solar densities. When the temperature at the bottom of the hydrogen layer becomes just high enough to start burning the hydrogen, nuclear energy is released. The layer heats up, the nuclear reactions work faster and the energy production increases, which results in a further increase of the temperature. Because of the high densities in the white dwarf gas the gas is degenerate, which means the gas pressure does not increase with increasing temperature. Therefore the hydrogen burning layer remains in hydrostatic equilibrium in spite of the increasing temperature, and the nuclear reactions proceed still faster leading to an exponential increase in temperature. If the temperature becomes high enough the degeneracy of the gas will be removed. (For each

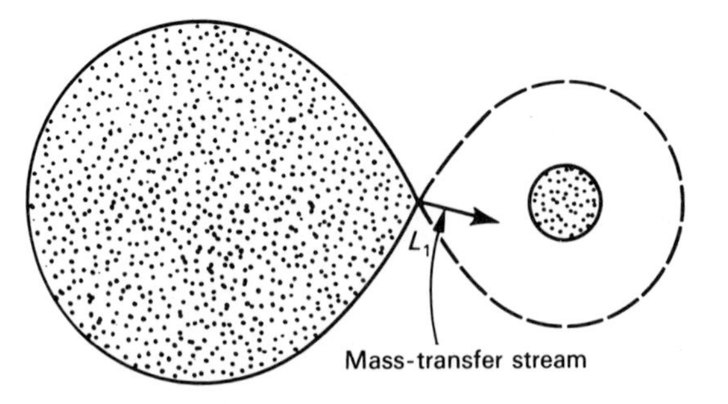

Fig. 17.21. The Roche lobe is shown for a binary system in which the larger left-hand star fills its Roche lobe. At the singular point L_1 mass can flow from the large star to the small one, which is much smaller than its section of the Roche lobe. (From Shu 1982.)

density there is a well-defined temperature for which this will happen.) Once this temperature is reached, the gas pressure increases again with increasing temperature and the pressure forces exceed the gravitational forces. The nuclear burning region expands and cools due to the expansion. The nuclear reactions slow down. A new equilibrium with a non-degenerate burning zone is set up. It is believed that the runaway burning in the accreted hydrogen and the subsequent expansion are the origin of the nova explosions. We must, however, admit that there are still serious problems for the theoreticians who actually want to reproduce a nova explosion on the computer. A large carbon over-abundance is required to make the nuclear reactions work fast enough and the chemical abundances observed in a nova explosion are not reproduced theoretically. Nevertheless this scenario seems to be the only viable one proposed so far.

18

Our sun

18.1 Introduction

The sun is, of course, the star nearest to us and is, therefore, the best-studied star. We have mentioned the sun several times as an example when we talked about distances of stars, effective temperatures and masses of stars, as well as angular radii. For all these studies we considered the sun to be just one of the normal stars, which it most probably is. It is the most thoroughly studied normal star. The sun is also the *only* star for which we can get high spatial resolution, which enables us to observe fine details on the surface which we will not be able to observe on other stars, at least not in the foreseeable future. These high spatial resolution studies of the sun reveal many features and processes which may well be also going on in other stars, but which we are not able to study in any other celestial object. Some of these features are the solar chromosphere and corona, though ultraviolet observations by means of satellites now permit us also to study global properties of these outer layers of stars other than the sun.

Another such phenomenon, which we can study in detail only in the sun, is the solar activity, which means flares, sunspots, and the whole solar cycle of activities. Again, observations with very high resolution and very sensitive receivers now permit us to study global effects of activity in other stars which seem to show activities similar to the sun. We would never know how to interpret these observations if we did not have the sun nearby so that we can study the solar surface in detail.

18.2 The surface of the sun

How does the sun look at higher spatial resolution? Not at all smooth and homogeneous, as one might expect for a hot gas ball.

First we see some large, dark spots, the so-called sunspots (see Fig. 18.1). We will talk about those later. Outside of the sunspots, we see a network called the solar granulation. It seems to be a network of dark structures with brighter blobs in between. According to what we discussed previously, dark means less radiation, which means a lower temperature. So we see regions with different temperatures on the solar surface. The spots are the coolest regions. The effective temperatures of the spots are about 3700–4500 K. For the granulation, the temperature difference between the dark and the bright region is only ~ 200–400 K, with an average temperature of about 5800 K, as we saw previously.

We will understand the solar granulation better after we have looked at a solar spectrum with very high spectral and spatial resolution (see Fig. 18.2)

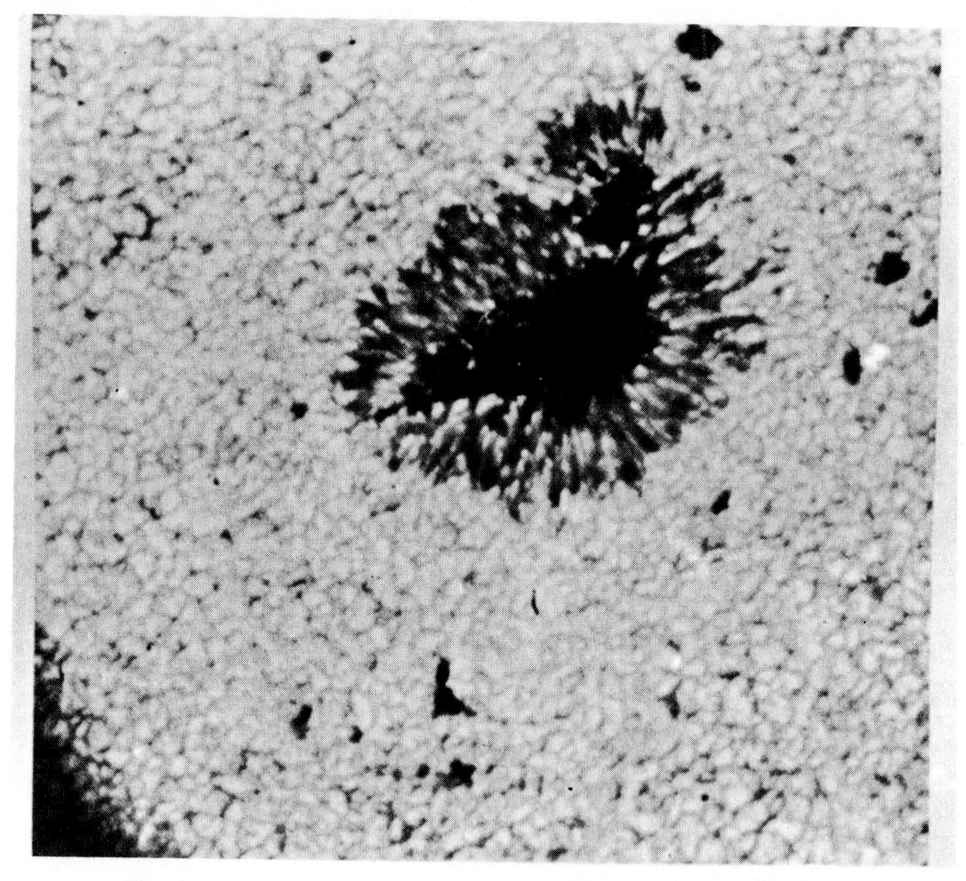

Fig. 18.1. A photograph of a small section of the sun is shown with very high spatial resolution. We see the dark umbra of a sunspot and the somewhat brighter penumbra around the umbra. In addition, we see small dark pores. The rest of the solar surface is not smooth at all; it is covered by a pattern of bright so-called 'granules' imbedded in the intergranular network. See also Fig. 18.4. (From Unsöld 1982; photograph by Schwarzschild using the stratospheric telescope.)

Because of the very high wavelength resolution, the spectral line is now rather broad. The interesting thing is that it is no longer a straight line but rather wiggly. Remember that in Section 9.2 we talked about the Doppler effect? That the spectral lines are shifted towards longer wavelengths when the light source is moving away from us, and towards shorter wavelengths when the light source is moving towards us? This is obviously what we see in these wiggly lines: some parts of the solar disk are moving towards us while other parts are moving away from us. Moving towards us means rising in the solar photosphere, and moving away from us means falling into the solar

Fig. 18.2. A narrow wavelength band of the solar spectrum is shown with very high spectral resolution and very high spatial resolution along the spectrograph slit (see Fig. 18.4). At such high spatial and spectral resolution the solar spectral lines are not at all straight lines. They are quite wiggly due to the Doppler shift, which is different for different spots of the sun seen at the different locations along the spectrograph slit. The continuous spectrum between the lines also looks streaky. It is brighter at the locations of the bright spots on the sun – the granules – and darker in between at the intergranular regions. When looking at the wiggles of the line profile we see that the Doppler shift goes preferentially to the left in the bright regions and preferentially to the right in the dark regions. In the photograph the wavelength increases toward the right. For the dark regions the velocities are therefore pointed away from us, while for the bright granules they are pointed toward us. (From Zirin 1966.)

photosphere (see Fig. 18.3). The spectrum in Fig. 18.2 was taken by having the slit of the spectrograph cover a small area on the solar disk, as shown in Fig. 18.4. Along the lengths of the slit we see the spectra of different elements of the granulation. At the point where we see a bright granulation element, we have more light in the spectrum; at places where we have a dark portion of the granulation pattern in the slit, we have less light in the spectrum.

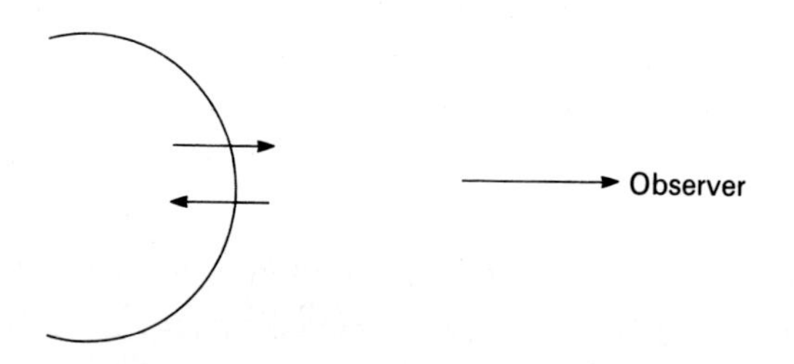

Fig. 18.3. Solar material moving towards us is rising in the solar photosphere, while material moving away from us is falling into the solar surface.

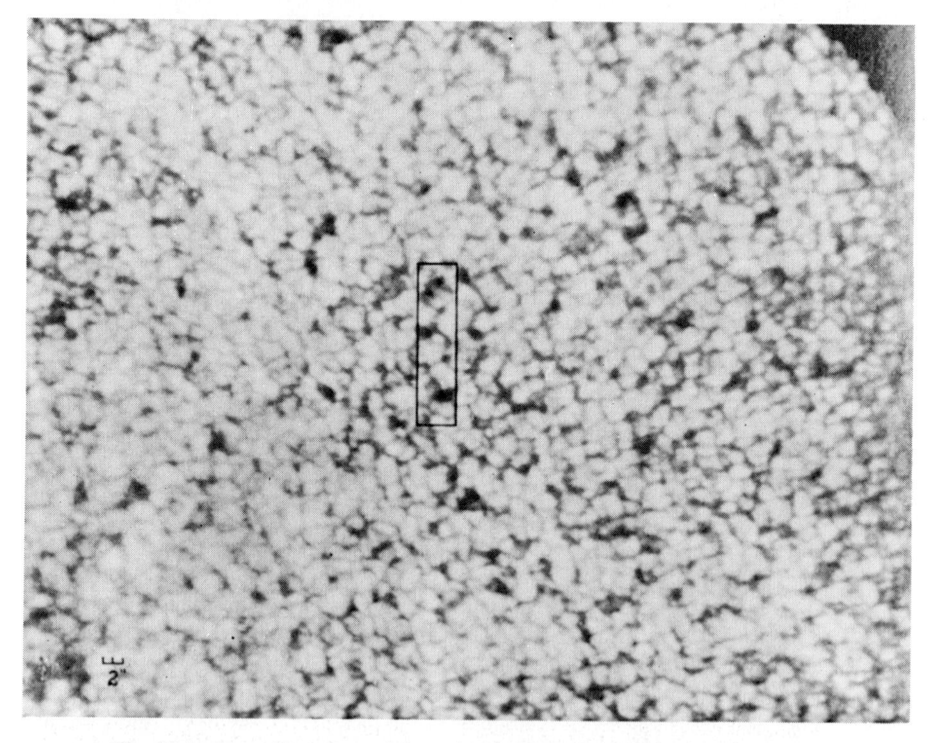

Fig. 18.4. A small section of the solar surface with the solar granulation is shown. Such an image of the sun is projected onto the spectrograph slit, which is outlined in the figure. Along the slit different bright and dark spots are seen. The spectrograph produces a spectrum of each spot along the slit. In this way a spectrum with high spatial resolution along the slit can be obtained, as seen in Fig. 18.2.

A careful look at the spectrum in Fig. 18.2 shows that bright streaks, showing the spectrum of a bright granule, usually show a blueshift of the lines i.e., a wiggle toward shorter wavelengths, while dark streaks, showing the spectrum of dark areas on the solar surface, show lines with wiggles towards longer wavelengths, at least in most cases.

The solar granulation then tells us that we have a pattern of upward and downward moving columns, the upward moving gas columns being brighter and therefore hotter than the downward moving ones. The diameters of these columns appear to be 500–700 km, but we have to remember that the turbulence in our Earth's atmosphere, limits our resolution to 1/2–1 second of arc. One second of arc corresponds to 700 km on the solar disk in the center of the sun. If the granulation has, indeed, a smaller size pattern, we are not able to resolve it. Satellite telescopes, which avoid the problem of seeing, so far, have had too small an aperture to resolve much smaller structures. The space telescope will be too light sensitive to be used on the sun which would burn it out.

18.3 The outer layers of the sun

18.3.1 The solar chromosphere

Before the time of rockets and satellites, the outer layers of the sun could only be observed during solar eclipses. The sun proper emits about 10 000 times more light than the chromosphere and one million times more than the corona, of which we show a picture in Fig. 18.5.

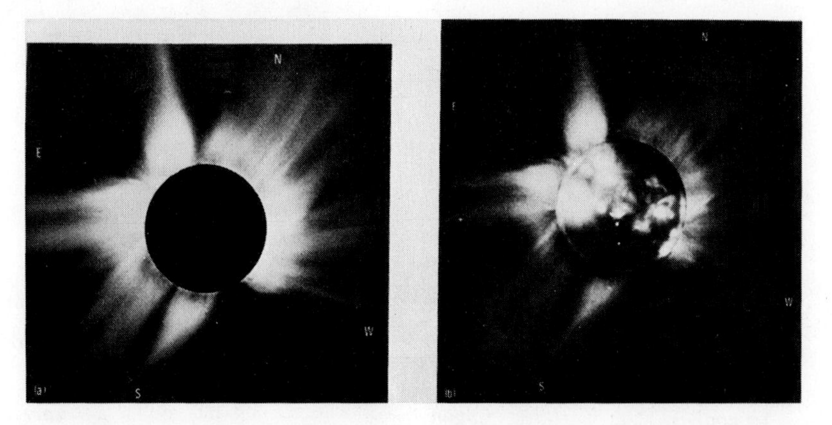

Fig. 18.5(*a*). The solar corona as seen during a solar eclipse. On the right-hand side a solar X-ray picture, taken shortly after the eclipse, is photomontaged on the shadow of the moon. X-ray emission is strong in regions of large solar streamers. (From Gibson 1973.)

The deepest of these outer layers becomes visible during solar eclipses just after second contact, when the solar disk proper is just covered up (see Fig. 18.6). This layer is called the chromosphere because it looks rather colorful, reddish. The spectrum of the chromosphere looks very different from that of the photosphere. The Fraunhofer lines turn into emission lines and the lines in the chromospheric spectrum are brighter than the neighboring continuum. The spectrum is also called the flash spectrum because it is visible for only a few seconds. In Fig. 18.7 we show such a flash spectrum. This spectrum was taken with a slitless spectrograph so we see an image of the solar limb for each emission line. (Usually we see an image of the slit of the spectrograph illuminated by the star which gives us spectral lines which are images of the spectrograph slit in the different colors.)

When the moon covers up more of the deeper chromospheric layers, the

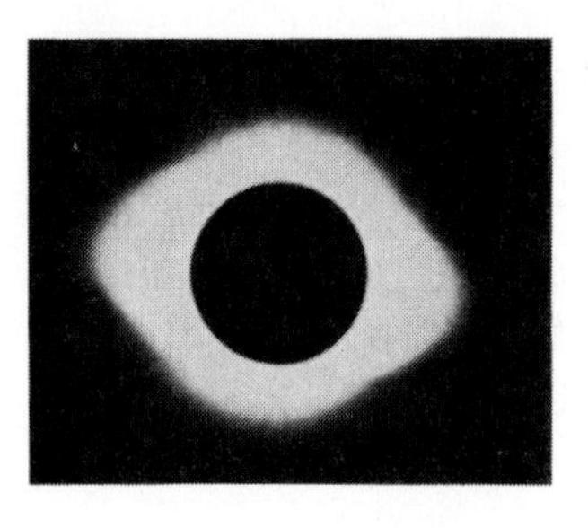

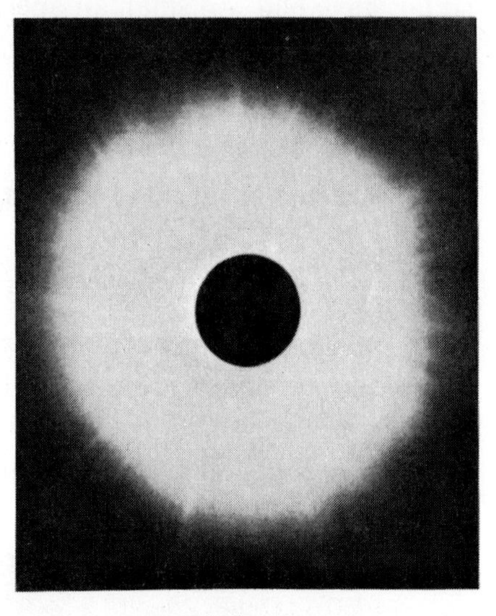

Fig. 18.5(*b*). The solar corona during sunspot minimum and maximum. (From Gibson 1973.)

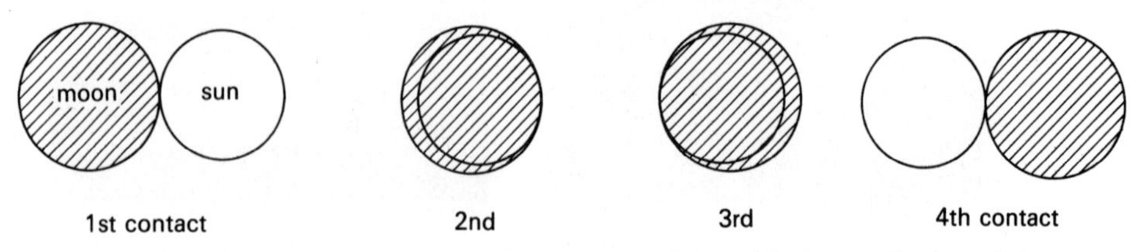

Fig. 18.6. We show the phases of a solar eclipse, with the numbering of the contacts given. The angular size difference between the sun and the moon is exaggerated.

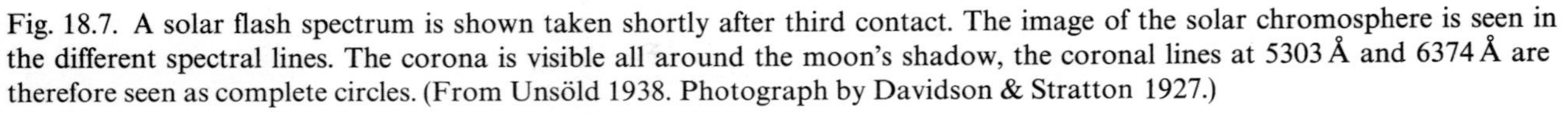

Fig. 18.7. A solar flash spectrum is shown taken shortly after third contact. The image of the solar chromosphere is seen in the different spectral lines. The corona is visible all around the moon's shadow, the coronal lines at 5303 Å and 6374 Å are therefore seen as complete circles. (From Unsöld 1938. Photograph by Davidson & Stratton 1927.)

higher, less luminous layers become visible. Lines of more highly ionized particles become visible, showing the increasing temperatures of the higher chromospheric layers. In fact, one can study the temperature and pressure stratification of the chromosphere by first studying the spectrum from the highest layers, observed last. Knowing what the emission is from these layers, we can subtract this contribution from the spectrum of the next deeper spectrum to get the emission from the deeper layer only, etc. Of course, we have to take into account that we always look tangentially along the solar limb. Our line of sight always cuts through all the layers (see Fig. 18.8), but knowing what the emission of the highest layers is, we can take that into account.

In Fig. 18.9 we show the temperature structure in the solar chromosphere and in the overlying transition layer, as derived from such studies of the solar chromosphere during solar eclipses. The temperature rises rather steeply in these layers until it reaches more than a million degrees in the corona. The region with temperatures between 30 000 K and 10^6 K is called the transition region (from the chromosphere to the corona).

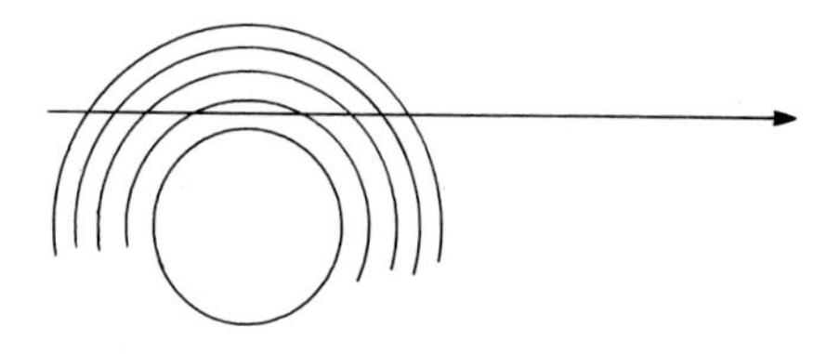

Fig. 18.8. When observing the solar chromosphere during a solar eclipse, our line of sight always cuts through all layers of the chromosphere and corona.

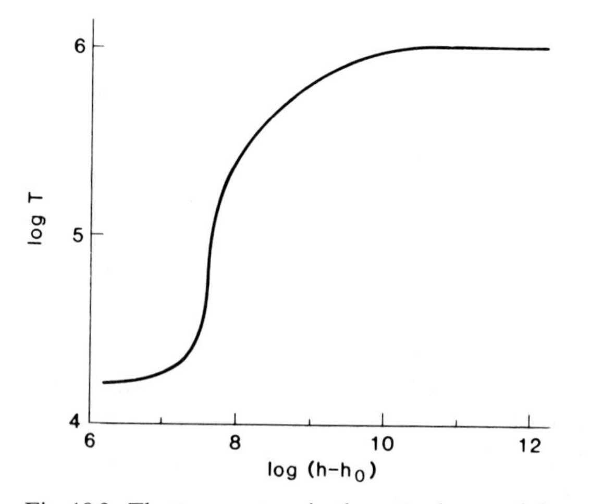

Fig. 18.9. The temperature in the outer layers of the sun is shown as a function of height above a reference level, h_0, called the 'surface' of the sun. (From Böhm-Vitense 1985.)

The sun is a rather cool star, therefore the photospheric light is quite dim at short wavelengths, which means in the ultraviolet. At these wavelengths the emission from the solar chromosphere is actually stronger than the light from the solar disk because the chromosphere has a higher temperature. This ultraviolet light can not be seen from the surface of the Earth because the atmosphere absorbs all the ultraviolet light, but from rockets and satellites it can be measured. We can therefore now also observe the chromospheric spectrum without an eclipse. This fact also enables us to observe chromospheres and transition layer emission from other stars for which previously no chromospheric or coronal light emission could possibly be observed. We did not know whether our sun was the only star to have a chromosphere and corona, or whether all cool stars have these high-temperature outer layers. We know now that cool main sequence stars with spectral types earlier than M5 have chromospheres and coronae, and all cool giants and the cool main sequence stars have at least chromospheres.

18.3.2 The solar corona

After second contact during solar eclipses, the chromospheric spectrum remains visible for about 20 s, after which time the whole chromosphere is also covered up by the moon and only the corona remains visible. Photographs of the solar corona show that it looks very different at different times (see Figs 18.5 and 18.6), depending mainly on the number of sunspots which are visible on the disk at that time, of course before or after the eclipse. During sunspot maximum the corona is large and looks almost circular, while during minimum it is only extended in the equatorial region, the region where the spots are mainly found. Large rays are seen showing us the directions of the magnetic field lines in the corona. In the equatorial regions the magnetic field lines show loops, while in the polar regions the magnetic field lines point radially outwards and the field lines look 'open.'

The spectrum of the corona shows spectral lines from ions which have lost many electrons, up to 20, indicating temperatures up to 2 or even 3 million degrees.

Such high temperature plasma also emits X-rays which can also be observed from rockets and satellites. We can actually take X-ray photographs of the sun, as seen in Fig. 18.10. This picture shows that the X-ray emission from different coronal regions is very different. It is generally larger in regions of large sunspot groups. There are large regions on the solar surface where the X-ray emission is so small that the X-ray photograph looks dark. These regions are called coronal holes. They have actually nothing to do with holes, but are just too cool or have too low densities to emit X-rays. They occur mainly in regions of 'open' field lines which reach very far out from the sun before they close, and from which matter can more easily escape from the solar surface than in regions with closed field lines, i.e. in the loops.

18.3.3 The energy source and the heating of the chromosphere and corona

We said earlier that heat transport can only be from higher temperature regions to cooler ones. Since the sun loses energy at the surface, which has to be replenished by heat transport from the deeper layers, the temperature in the sun has to decrease outwards. We cited the absorption line spectrum as proof of such a temperature gradient outwards. How then can the temperature increase in the very outer layers of the sun? How can the solar radiative heat transport be from the relatively cool photosphere to the high temperature corona? The answer to the latter question is that the heat does not flow *to* the high temperature corona, it only passes *through* it. The hot corona contains so little material that the photospheric radiation does not notice the existence of this material. There is essentially no absorption of photospheric radiation in the corona. The optical depth of the corona is of

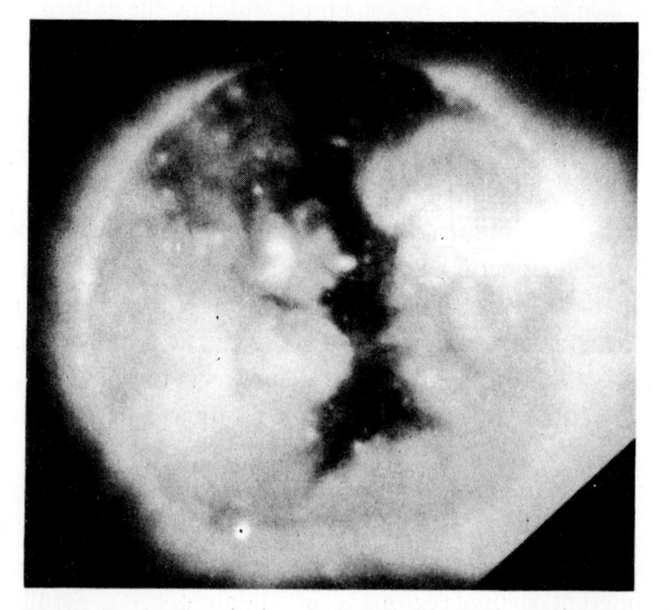

Fig. 18.10. An X-ray image of the sun showing a coronal hole extending right across the solar surface. In this region the X-ray emission is very low. (Photo: NASA, courtesy of L. Golub and Harvard-Smithsonian Center for Astrophysics. From Gibson 1973.)

the order of 10^{-10}. The average optical depth of the chromosphere is about 10^{-4}. These layers do not get their energy from the radiation which comes from the deeper layers and is absorbed in these layers, as is the case for the solar photosphere. What then is the heating mechanism for the chromospheres and coronae? This question has stimulated an intense study, especially during the last decade when the International Ultraviolet Explorer (IUE) satellite and the X-ray satellite, the Einstein Observatory, were launched. The answer to the question for the specific sources of energy in the different layers can not yet be given with certainty; the discussions are still going on.

I think it is generally agreed that the ultimate energy source is the so-called hydrogen convection zone which is situated just below the visible layers of the photosphere. We saw the granulation picture. The Doppler shifts of the wiggly lines show velocities of the order of 1–2 km/s, which means about 5000 km/h. These velocities are generated in the hydrogen convection zone (see Volume 2). In these strong velocity fields, much noise is generated, which means acoustic waves, which in the low density chromospheric layers develop into shock waves. They may also lead to the generation of magnetohydrodynamic waves in the general solar magnetic fields and especially in the strong magnetic fields around the sunspots.

These forms of energy are not heat energy, but are in brief called 'mechanical energy.' Waves are influenced by the temperature structure, but they can propagate in any direction regardless of the temperature gradient. Their propagation is mainly influenced by the density gradient and by the magnetic fields. The acoustic waves may dissipate some energy in the low chromospheric layers. When they develop into shock waves they dissipate their energy much faster. Magnetohydrodynamic waves cannot dissipate their energy as fast in the chromosphere. They may be the vehicle by which mechanical energy can be transported into the corona where they may be partly transformed into waves similar to acoustic waves which may possibly heat the corona, though other heating mechanisms are also possible. We do not yet completely understand the details of these wave generations and dissipations, but we do know that there is enough energy in these waves to replenish the energy loss in the chromosphere and corona of the sun and similar stars. As in the photosphere, the resulting temperature stratification is determined by a balance between the energy input and the energy loss. In the high layers coronal temperatures around 1 or 2 million degrees are reached. These high temperature regions are the origin of the so-called 'solar wind'. The solar wind energy increases steeply with increasing temperature of the corona. It takes out this energy and is therefore a very efficient cooling mechanism.

18.3.4 The solar wind

The solar wind was first discovered by L. Biermann who studied the tails of comets. The ionized parts of these tails are always pointed away from the sun, as if something pushes them out. He calculated that the radiation of the sun alone was not strong enough to do it, so there had to be some material flying away from the sun at high speeds, pushing the comet tail out. When rockets studied the motions and magnetic fields in the Earth's magnetosphere, they found that the magnetic field is deformed by a stream of matter coming from the sun at high speeds, 500 km/s. This stream got the name 'solar wind.' The speeds are really more than a wind, but the stream has a very low density, $\sim 10^{-23}\,\mathrm{g\,cm^{-3}}$.

In solar surface regions with closed magnetic field lines the solar wind has to fight the magnetic forces. Ionized material cannot flow across magnetic lines of force. (Charged particles spiral around the lines of force). The solar wind cooling is therefore less efficient in regions with closed lines of force, where the temperatures therefore remain high. In regions of 'open' field lines the cooling by the wind is much more efficient and the temperature is expected to be lower in such regions of the stellar surface. This explains the existence of coronal holes, the regions of low X-ray emission in the corona.

18.4 The active sun

18.4.1 The Butterfly diagram

In a previous photograph of the sun (Fig. 18.1) we saw a dark sunspot. Each of these sunspots lasts for about a month and then disappears while new ones show up. The number of sunspots, and their position, varies with time.

The following diagram (Fig. 18.11) called the 'Butterfly' diagram shows the position of sunspots on the sun as a function of time. The ordinate is the latitude on the sun. The number of sunspots shows a cycle of approximately 11 years duration. For a new cycle, a few spots show up first at rather high latitudes, $\sim 40°$, and the next new spots occur closer to the equator. The sunspots themselves do not move toward the equator, but the zone does.

While the last spots from a given cycle are still present near the equator, the first spots of the new cycle occur at higher latitudes.

What are these dark spots? Why are they dark? The first insight was gained when Hale found in 1908 that the sunspots were regions of large magnetic fields.

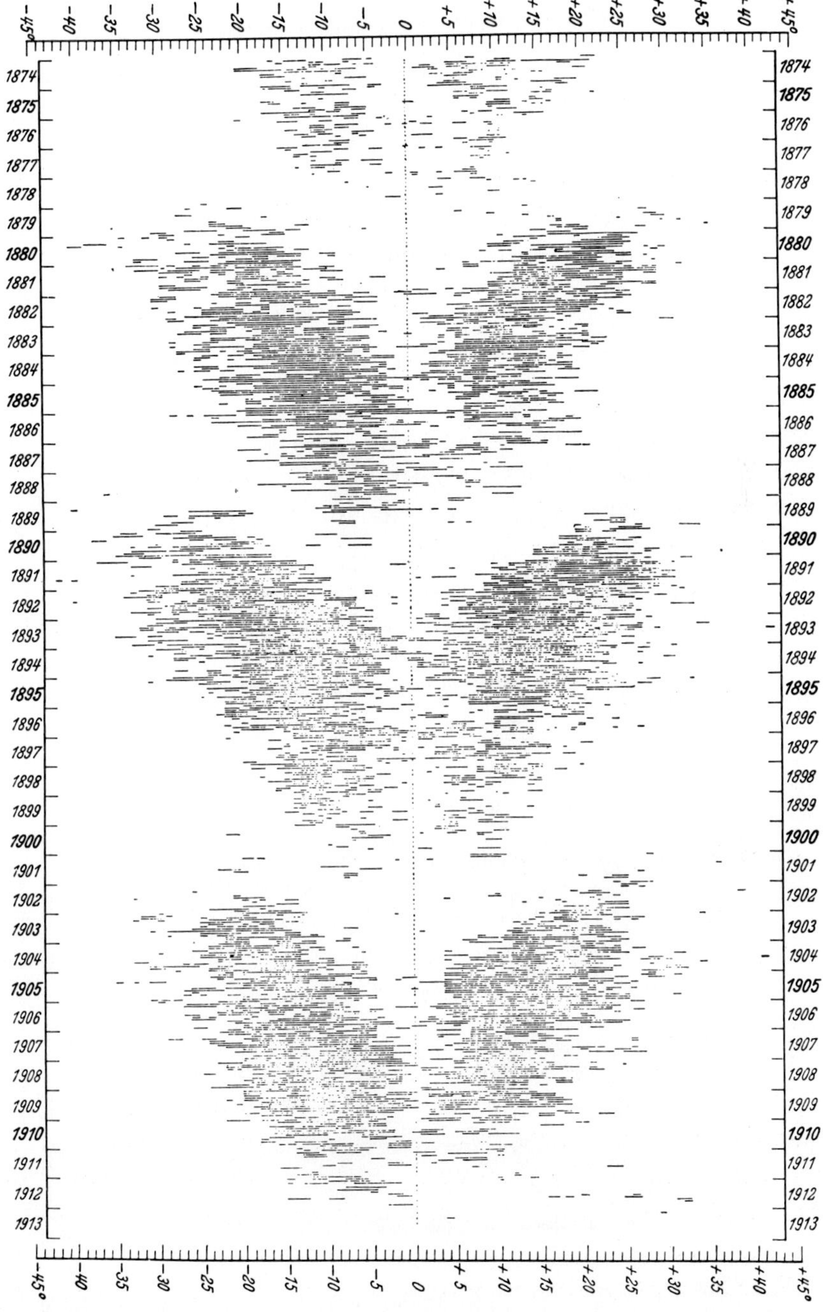

Fig. 18.11. The solar latitude of sunspot positions is shown as a function of time. The first sunspots of a new cycle occur at latitudes around 40°. Later in the cycle sunspots occur at lower latitudes. The last sunspots of a cycle occur rather close to the equator, while the first spots of the new cycle appear already at latitudes around 35°. Because of its appearance this diagram is called the butterfly diagram. (From Unsöld 1938.)

Spectrograph slit

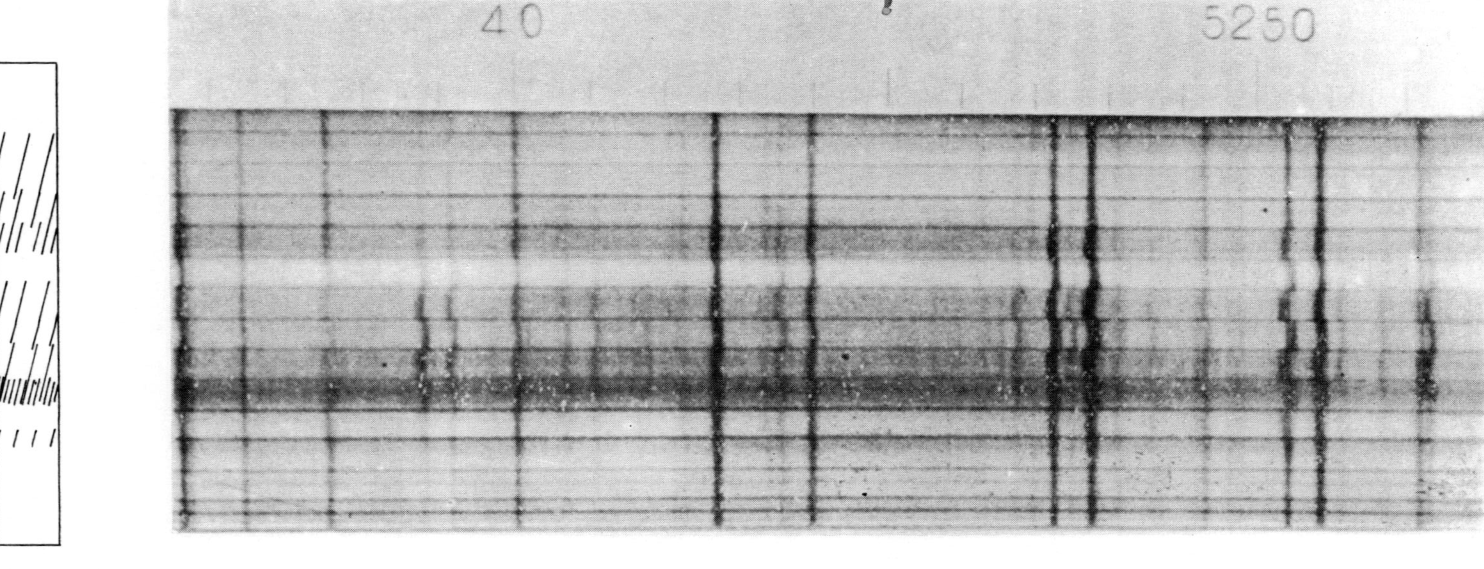

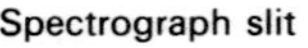

Fig. 18.12. Sunspot spectra obtained with the magnetograph are shown. The vertical direction is the space coordinate along the spectrograph slit. The dark regions show the position of the sunspot. The horizontal coordinate is the wavelength of the light. The light seen in two adjacent spectra comes from the same area on the sun but is of opposite direction of circular polarization (see Chapter 16). The Zeeman effect due to the magnitude field in the sunspots is clearly seen in the shift of the spectral lines of the spectra of opposite direction of polarization, indicating magnetic fields of several thousands Gauss. (From Kiepenheuer 1953.)

18.4.2 *The magnetic fields in sunspots*

In Section 15.2 we discussed how we can measure magnetic fields in stars by means of a magnetograph first constructed by Babcock. This instrument makes use of the Zeeman effect: if a light source is put into a strong magnetic field, the spectral lines split up in several components, the amount of splitting increasing with the strength of the magnetic field. The wavelength difference between the various components of a line can then be measured and the field strength determined. The direction of the field can also be determined because the pattern of the components, i.e., their relative strengths and the way the light is polarized, changes with the direction of the field relative to the line of sight.

Measurements of the Zeeman effect in sunspots revealed that sunspots have magnetic fields of several thousand Gauss in the darkest part which is called the umbra (see Fig. 18.1). In Fig. 18.12 we see a spectrum of a sunspot in which one strip shows the spectrum of one direction of polarization, and the adjacent strip shows the other direction of polarization. The next pair of spectra shows the adjacent area of the sunspot. The line shifts of two adjacent spectra clearly show the presence of a magnetic field.

The field appears to be mainly longitudinal, which means vertical in the umbra, but curves around in the penumbra. We do not understand at all why there is such a sharp boundary between umbra and penumbra and the surrounding area, nor do we understand the difference in brightness between umbra and penumbra.

When studying the polarity of the magnetic fields, it is found that whenever there are two spots closely associated, they have opposite polarities. In large sunspot groups, one also finds opposite polarities in the two halves of the group. The interesting thing is that during one sunspot cycle the north

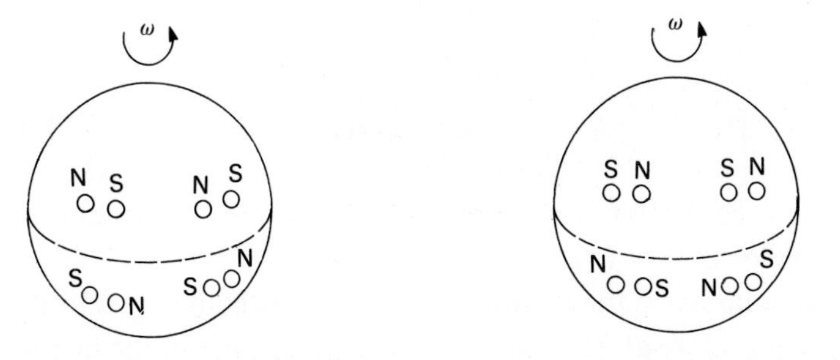

Fig. 18.13. During one 11-year sunspot cycle the magnetic north polarity is always in the preceding spot on the northern hemisphere and in the following spot in the southern hemisphere. In the next 11-year cycle the polarities are reversed.

polarity is always on the preceding spot (preceding with respect to the solar rotation) and the south polarity on the following spot for one hemisphere and opposite for the other hemisphere. During the next cycle the polarities are reversed (see Fig. 18.13). In reality, the sunspot cycle is therefore ~ 22 years if we include the magnetic polarity variations.

18.4.3 Differential rotation of the sun

The rotation rate of the sun can best be measured by measuring the sunspot velocity across the disk (see Fig. 18.14). The average rotation period of the sun is nearly 26^d, almost identical to the orbital period of the moon. However, the spots closer to the equator rotate somewhat *faster* than those further away. This is called the differential rotation of the sun.

18.4.4 Possible explanation of the sunspot cycle

At solar temperatures and densities several of the heavy elements like iron and silicon lose one electron, therefore the solar material contains a large number of free electrons. This makes it a very good conductor for electrical currents. Such gas with charged particles is also called a plasma. A plasma can not move through magnetic lines of force because the electrons always have to spiral around the magnetic lines of force. If a plasma wants to move perpendicularly to a magnetic field it has to take the magnetic lines of force along. We say the magnetic lines of force are frozen into the material.

Based on this information we can give the following very simplified explanation of the solar cycle.

We have just seen that the solar equator is rotating faster than the higher latitudes. Due to this differential rotation, the magnetic lines of force are wound up as shown in Fig. 18.15. The direction of the lines of force is reversed in the northern and southern hemispheres.

If such a magnetic 'rope' breaks through the solar surface in different regions, pairs of magnetic spots are seen for which the direction of polarity is always the same (see Fig. 18.16). Since in the northern and southern hemisphere the field directions along the 'ropes' are opposite, the north–south polarity of the spot pairs is also reversed.

In the next solar cycle the overall solar magnetic field is reversed and the polarities are therefore all reversed but remain opposite in the northern and southern hemispheres.

We want to emphasize that this is a very simplified picture of the processes involved, but we can at least understand the main features in this way. To do it better we have to apply the dynamo theory which for non-rigid rotation gives periodic changes of the overall magnetic field.

18.4.5 *The temperatures of the sunspots*

Since the sunspots are dark, the gas in these magnetic regions must be cooler than in the surroundings. What makes the sunspots cool? To cool

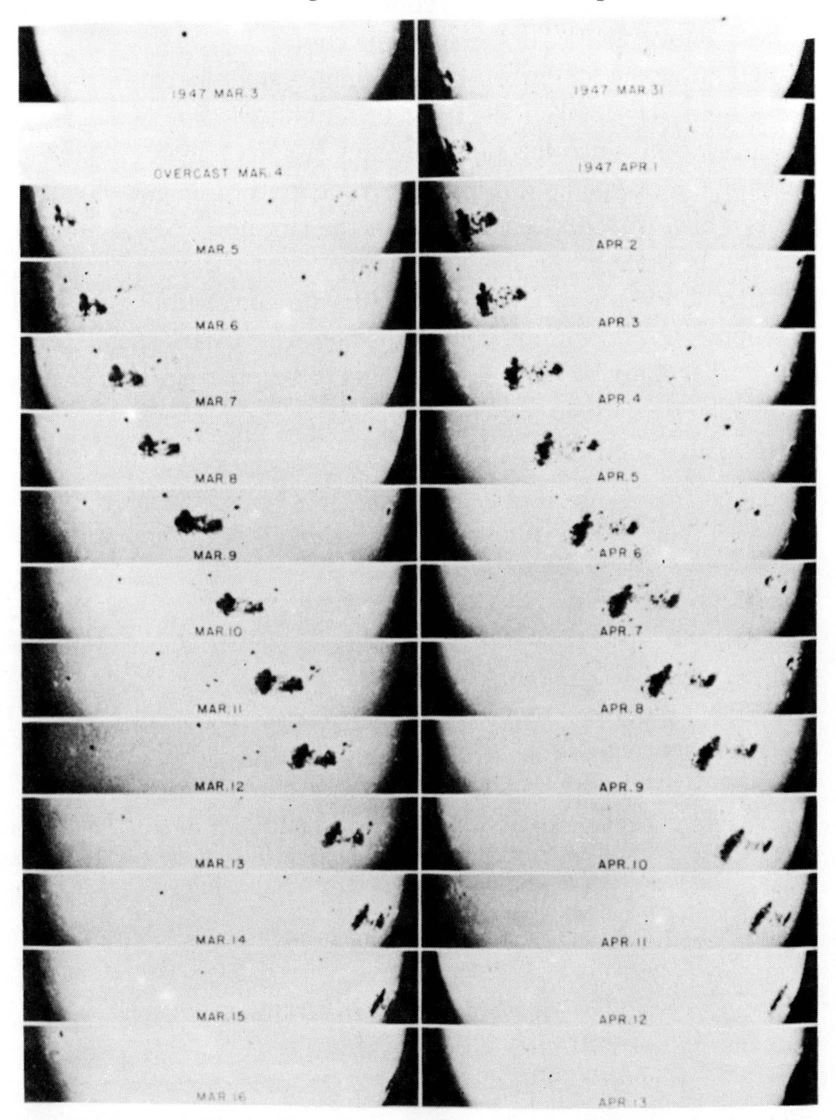

Fig. 18.14. The rotation rate of the sun can be seen in the motions of the sunspots around the sun. Spots close to the equator move somewhat faster than those at larger latitudes. An average rotation period of 28 days is found. (From Abell 1982.)

the gas we must either take out more energy at the surface or put in less energy at the bottom. Astronomers cannot quite agree on the correct explanation. The suggestion by Biermann (1941) was that less energy is being fed into the sunspot gas at the bottom. We said earlier that below the surface of the sun we find a convection zone. The moving material transports energy to the surface. In the strong magnetic fields of the sunspot, the material cannot move easily, so the convection probably transports less energy to the surface. That could leave the sunspots cooler.

On the other hand, Parker (1974) maintains that this cannot be the correct explanation. He thinks there must be additional cooling. It cannot be by light radiation, since the sunspots emit less light than the surroundings. Perhaps the magnetic field leads to some kind of magnetohydrodynamic waves which transport energy out of the sunspots? Much more research needs to be done in this field.

If Bierman is correct, then only stars with convection zones could have dark spots. If stars without convection zones have dark spots, then it seems that Parker must be right. There seems to be observational evidence that

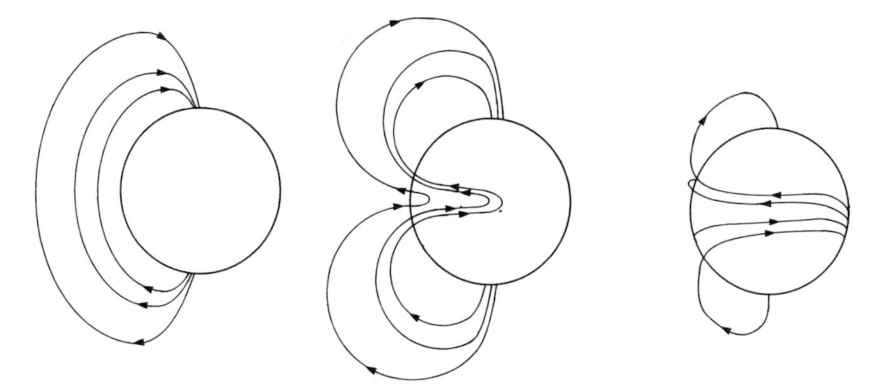

Fig. 18.15. The magnetic lines of force on the sun are 'frozen' into the material. They are wound up due to the differential rotation of the sun. The horizontal components of the magnetic lines of force are then in opposite directions on the northern and southern hemispheres.

Fig. 18.16. When the magnetic ropes seen in Fig. 18.15 break through the solar surface, sunspot pairs of opposite polarities are seen, always with the same orientation in one hemisphere but opposite in the southern and northern hemisphere.

stars without convection zones may also have dark spots. It is not known whether those might be magnetic spots.

18.4.6 *Stellar activity and stellar rotation*

We saw in Chapter 13 that stellar rotation rates decrease abruptly for stars cooler than spectral type F5. For these cooler stars the rotational velocities are in fact so small that they cannot be measured from the line widths anymore. It appears that we can measure the rotation rate of these stars from starspot activities similar to the one observed for the sun. In active solar regions where a large number of spots are seen, we also see in the center of the strong Ca^+ lines at 3933 and 3968 Å a weak emission line. This line is stronger in active regions than for the overall solar spectrum. When, due to solar rotation, these active regions rotate to the back side of the sun where we cannot see them, then in the overall spectrum of the sun the emission in the center of the Ca^+ lines becomes weaker. For the sun this effect is difficult to measure because the emission is so weak, but for many cool stars this emission is much stronger.

When we measure the stellar Ca^+ K line emission at 3933 Å as a function of time, we can frequently see variations which are semi-periodic. This is understandable if the star is rotating. When the activity region disappear behind the star, the emission is reduced; when the spots appear again on the other limb, half a rotation later, the emission becomes strong again. When the spot finally disappears, the emission from this region cannot be seen any more, but in the meantime there may be another active spot group at another place on the star for which the emission can be observed. But this group may already be further advanced on the stellar disk, so there will be a phase shift between the rotation of the first spot and the one for the second spot. This causes deviations from strictly periodic behavior. Nevertheless, rotation periods have been measured in this way. It is found that stellar chromospheric activity generally increases with increasing rotation rates of the stars. Only early F stars seem to be an exception, the reason for this is not yet understood.

19

Interstellar absorption

19.1 Introduction

Even though this volume is dedicated only to stellar observations and the theory of stellar structure, we have to talk a little about the interstellar material, which means the material which is between the stars, because the light of the stars, especially of those which are very far away, has to pass through this interstellar material before it reaches us. Just as in the Earth's atmosphere, the star light is influenced by interstellar absorption, also called interstellar extinction. We cannot observe all the light which is emitted from the star because the interstellar gas and the interstellar grains, called dust, have absorbed part of the light. Just as for the Earth's atmosphere, we have to correct for interstellar absorption. For the Earth's atmosphere we can determine the extinction if we observe the star at different zenith distances, which means for different path lengths, through the atmosphere. For interstellar extinction we cannot do this since the path length through the interstellar medium is always the same. We have to find other ways to determine the influence of the interstellar extinction.

Basically there are two components of interstellar extinction: the absorption by the grains and the absorption by the gas. These components are usually, but not necessarily, associated. In the following paragraphs, we will discuss both components and see how we can find out about interstellar absorption.

19.2 The interstellar dust

19.2.1 General appearance

When looking at photographs of the Milky Way, we see regions with a large number of stars and, right next to them, regions where there are hardly any stars.

Dark patches intermixed with bright matter can be seen in many parts of the sky (see, for instance, Fig. 19.1). On the small-scale photograph of the Milky Way seen in Fig. 19.2 a dark lane cuts the Milky Way almost into two halves.

The question is, are the stars distributed so inhomogeneously, or is there something covering up the stars? When we try to find external galaxies, we

Fig. 19.1. In the sky, dark regions with very few stars are seen next to brighter regions with very many stars. In this picture we see the so-called Horsehead Nebula. (From Abell 1982.)

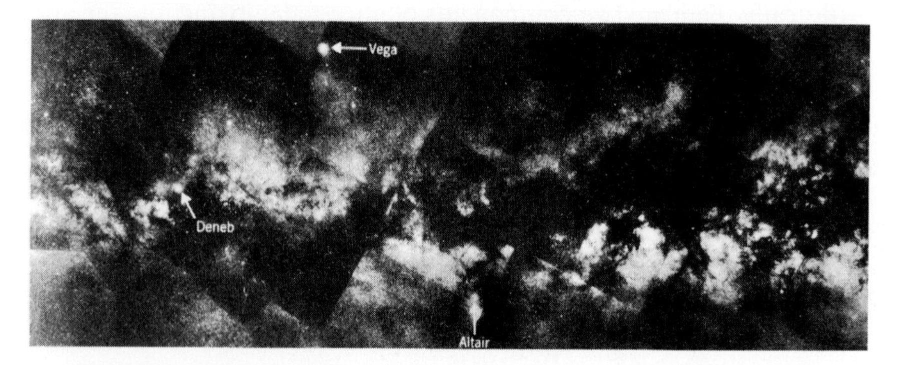

Fig. 19.2. A composite photograph of our Milky Way is shown. The border lines for different photographic plates can clearly be seen. We also see large dark lanes cutting through the Milky Way almost cutting it in two halves. (From Shu 1982.)

have trouble finding any in the plane of the Milky Way, which means in the plane of our Galaxy. For this reason the plane of the Galaxy, or of the Milky Way, is called the zone of avoidance. This strongly suggests that there is indeed absorption in the plane of the Galaxy because the external galaxies have no reason to care about the plane of our Galaxy. How can we decide with certainty whether the dark regions are indeed regions of strong absorption and not holes in the Milky Way? We can find out by statistics, by means of the so-called Wolf diagram.

19.2.2 The Wolf diagram

Let us look at a certain region in the sky and thereby observe all the stars in a given solid cone with an opening $\Delta\omega$. All these stars are projected against the given area in the sky (see Fig. 19.3). We study only stars of a given kind, say only A0V stars or, in other words, stars with spectra, like Vega. All such stars have the same absolute magnitude, M_v. (There are more luminous A0 stars around like giants and supergiants, but we saw that they can be recognized from their spectra.)

Let us assume that the stars are distributed homogeneously in this cone and that the number of A0V stars per pc^3 is n. In any volume V, the number of A0V stars is then $N = n \cdot V$. We now consider the part of the spherical shell at distance d with thickness Δd, which is cut out by the cone (see Fig. 19.3). The whole sphere is described by a solid angle 4π. The volume of the part of the sphere with solid angle opening $\Delta\omega$ is therefore given by a fraction $\Delta\omega/4\pi$, which means

$$V = 4\pi d^2 \, \Delta d \, \Delta\omega/4\pi. \tag{19.1}$$

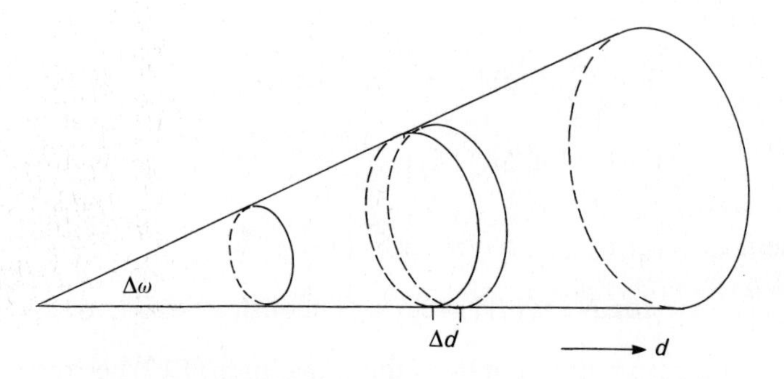

Fig. 19.3. Observing all the stars in a given area in the sky corresponds to observing all stars within a cone with an opening of solid angle $\Delta\omega$, whose size is determined by the extent of the observed region in the sky.

The number of A0V stars in this volume is then

$$N(d) = n \cdot V = n \cdot d^3 \cdot \Delta\omega \cdot \frac{\Delta d}{d}. \tag{19.2}$$

These stars all have an apparent visual magnitude

$$m_v = M_v + 5 \log d - 5, \tag{19.3}$$

or, for a change in distance by $\Delta \log d$ we find a change in apparent visual magnitude by

$$\Delta m_V = 5 \cdot \Delta \log d = 5 \cdot \log e \cdot \Delta \ln d = 2.171 \cdot \Delta \ln d = 2.171 \cdot \Delta d/d. \tag{19.4}$$

(A given Δm_V corresponds to a given $\Delta \ln d = \Delta d/d = \Delta m_V/2.171$.) For a given m_V we have then, according to (19.3)

$$\log d = \frac{m_v - M_v + 5}{5} = \frac{m_v}{5} - \frac{M_v}{5} + 1, \tag{19.5}$$

where d is obtained in pc. M_v is known from nearby A0V stars like Vega; m_v is measured.

The number of A0V stars with apparent magnitudes between m_V and $m_V + \Delta m_V$ is given by the number of stars with distances between d and $d + \Delta d$, for which the relation between Δm_V and $\Delta d/d$ is obtained from (19.4). From (19.2) we find

$$\log N[m_v(d)] = \log n + 3 \cdot \log d + \log(\Delta\omega \cdot \Delta \ln d), \tag{19.6}$$

if we assume a homogeneous stellar density with n stars per unit volume, which means $n = \text{constant}$. For a given choice of $\Delta \ln d$ corresponding to a chosen Δm_V for the magnitude interval for which we count the stars, all terms on the right-hand side of (19.6) are constant except for the $\log d$ term, which depends on the m_V for which we count the stars. We can therefore rewrite (19.6) in the form

$$\log N[m_v(d)] = 3 \cdot \log d + \text{const}, \tag{19.7}$$

with

$$\text{const} = \log \Delta\omega + \log \frac{\Delta m_v}{2.171} + \log n.$$

Replacement of $\log d$ by means of (19.5) now yields

$$\log N[m_v(d)] = 0.6\, m_v + \text{Const}, \tag{19.7a}$$

where the new Constant is different from the constant in (19.7) because it also includes the term stemming from the absolute magnitudes of the stars studied.

In a plot of $\log N$ versus m_V we should therefore find a straight line (see Fig. 19.4). This is the Wolf diagram.

Assume now that there is an absorbing cloud at distance $d = d_0$, for which $m_V = m_{V0}$. Suppose this cloud absorbs 75% of the light of the stars behind it, leaving 25% of the light, which means decreasing the light by a factor of 4, corresponding to an increase in magnitude by 1.5 for all stars behind the cloud. For a given distance, of course, the number of stars would not change; the $N(d)$ for $d > d_0$ would remain the same, but the apparent magnitude of these stars would be increased by 1.5. The relation between $\log N$ and m_v would then be as shown by the dashed line in Fig. 19.4. For the figure we also assumed that the cloud would extend over a distance Δd, corresponding to $\Delta m_v = m_{v1} - m_{v0}$. Without this finite extent to the cloud, the short, dashed line in the figure would have to be plotted horizontally because the m_v of the stars would either be changed by 1.5 or not at all. If the cloud is extended, the stars in the cloud have intermediate shifts, depending on their position in the cloud.

If at a larger distance $d = d_1$ there should be another cloud, we may find another shift in the distribution, as indicated by the dash–dotted curve in Fig. 19.4. Such a diagram can therefore not only tell us whether there is a cloud or not, but also where it is, how extended it is, and how much it absorbs.

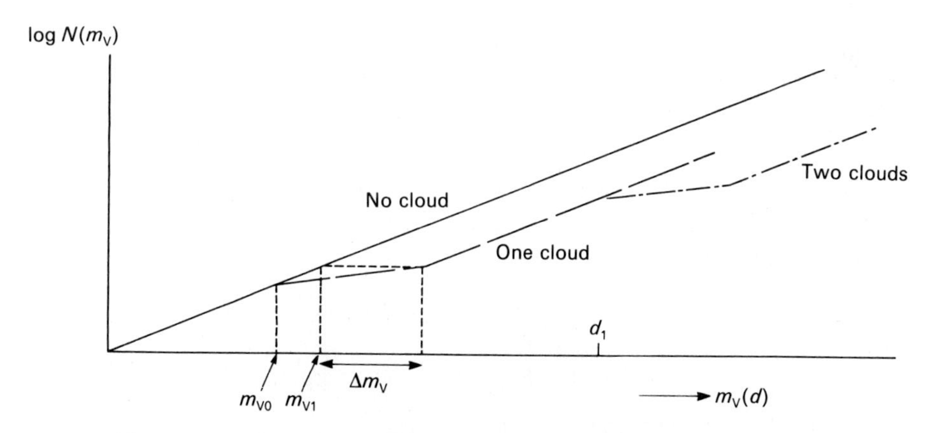

Fig. 19.4. A schematic Wolf diagram is shown. For homogeneous steller densities we expect a linear relation between the log of the number of stars of a given kind, log N (with a given absolute visual magnitude) and the apparent visual magnitudes (see the solid line). If there is an absorbing intersteller cloud between us and the stars the apparent magnitudes of the stars behind the cloud are all increased by an amount determined by the absorption in the cloud. The relation between $\log N$ and m_v is shifted as shown by the long dashed line. A second cloud behind the first one may give an additional shift (dash–dotted line).

What would we see if there was not an absorbing cloud, but instead there was a 'hole' in the star distribution, say, between the distances d_2 and d_3? If these distances correspond to apparent magnitudes m_{v2} and m_{v3}, then all the stars with $m_{v2} < m_v < m_{v3}$ would be missing. For these magnitudes N would be zero. The stars at larger distances would generally not be affected, but would probably have the same average density. The star counts for larger m_v would therefore be the same. The relation between log N and m_v would then look like the one shown in Fig. 19.5, where we have assumed that two holes exist: one at $m_v = m_{v2}$ and another one at $m_v = m_{v4}$. In this case the Wolf diagram looks very different from the one with the clouds. Such a Wolf diagram therefore permits us to distinguish between 'holes' in the stellar distribution and an absorbing interstellar cloud. In Fig. 19.6 we show an example of an observed diagram, clearly showing that interstellar clouds are obscuring the stars.

19.2.3 Interstellar reddening

The study described in the preceding section can, of course, also be done with the blue and ultraviolet magnitudes in the same way. We can then find out how much the same clouds absorb in the blue or in the violet. Such studies have shown that the absorption is larger in the blue and still larger in the ultraviolet. For most clouds we find

$$\Delta m_B = 1.3 \cdot \Delta m_v \quad \text{and} \quad \Delta m_U = 1.53\, \Delta m_v. \tag{19.8}$$

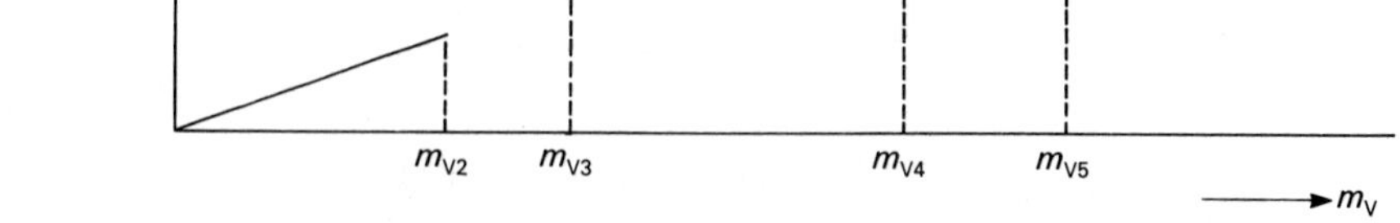

Fig. 19.5. We show the Wolf diagram expected if there are holes in the number density of stars at distances corresponding to apparent visual magnitudes m_{V2} and m_{V4}, but which would be normal again for distances corresponding to apparent visual magnitude m_{V3} and m_{V5}, respectively. In this case we would see no stars with apparent magnitudes between m_{V2} and m_{V3} or with magnitudes between m_{V4} and m_{V5}. There would be 'holes' in the log $N(m_V)$ plot.

Since the B and U magnitudes are more increased than the V magnitudes, the colors of the stars change due to the interstellar absorption. We find, roughly,

$$\Delta m_B - \Delta m_v = \Delta(B-V) = \Delta m_v(1.31-1) = \Delta m_v \cdot 0.31, \qquad (19.9)$$

or

$$\Delta m_v = 3.2 \cdot \Delta(B-V) \qquad (19.10)$$

or

$$\frac{\Delta(B-V)}{\Delta m_v} = 0.31 \quad \text{and} \quad \frac{\Delta(U-B)}{\Delta m_v} = 0.22 \qquad (19.11)$$

$$\Delta(U-B)/\Delta(B-V) = 0.72, \qquad (19.12)$$

but small differences may exist between different clouds.

In order to find the relation between the color change $\Delta(B-V)$ or $\Delta(U-B)$, we can, of course, also directly compare the colors of the A0V stars

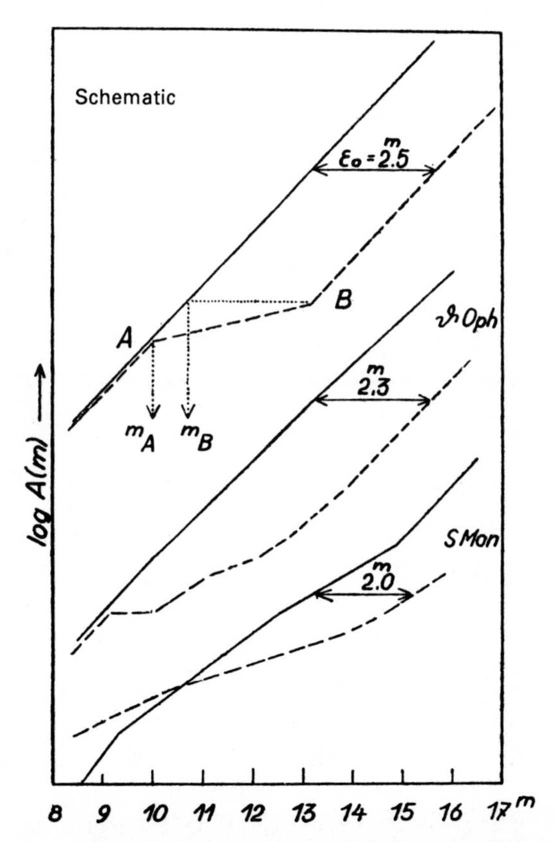

Fig. 19.6. Below the schematic Wolf diagram, already shown in Fig. 19.4, we see actually observed Wolf diagrams for a region around the star ϑ Ophiuchi and around the star S Monocerotis. The stars behind the cloud are dimmer by 2.3 and 2.0 magnitudes due to the absorption by the interstellar material in the clouds. (From Becker 1950).

in front of the cloud with those behind the cloud and determine the color changes and the ratio of Δ(B − V) to Δ(U − B).

In the far infrared the interstellar absorption becomes very small, actually close to zero. If we call a far infrared magnitude m_{in}, then the color change $\Delta(m_V - m_{in})$ for stars behind the cloud gives directly the change in visual magnitude for the stars behind the cloud. In this way the ratio of Δm_V to Δ(B − V), called R, is determined. It usually turns out to be $R = 3.2$, but may occasionally have values around 5 or perhaps even 6.

The Δ(B − V) and Δ(U − B) due to interstellar absorption are usually called E(B − V) and E(U − B), where E stands for color excess.

Once we have established that we do indeed have absorption in the interstellar medium, and once we know that the interstellar medium causes reddening, we can use this reddening effect to correct for the interstellar absorption. If we know what the true U − B and B − V colors are for unreddened stars, we can plot the two-color diagram for these true colors (see Fig. 19.7). In this figure we have also plotted the observed colors for some stars. For a given U − B the reddened stars have a much larger B − V. We know now that for any change in B − V by a value E(B − V) the change in U − B is E(U − B) with E(U − B) = 0.72 × E(B − V). This relation defines a direction in the two-color plot. In Fig. 19.7 we have indicated the direction in which the stellar colors move in the two-color diagram due to interstellar reddening. For an E(B − V) = 0.3, star D would appear in the position of star C. We can then trace backwards from where the star in position C came in the two-color diagram before it was reddened by tracing the observed colors

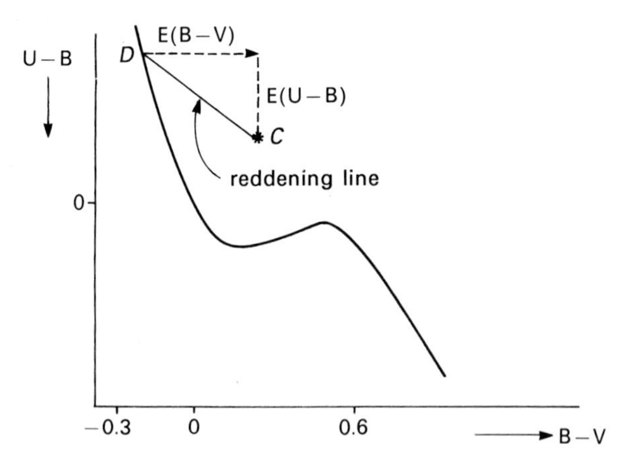

Fig. 19.7. The two-color diagram for the true colors of the stars is shown schematically by the solid line. Also shown is a point (*) for the observed colors of a star C. The direction of the reddening line E(U − B)/E(B − V) = 0.72 is also drawn in. Following the direction to the 'true' color line of normal stars tells us what the unreddened colors are of the observed star (point D).

back along the reddening line to the true color line. This gives us the true colors of the star. Since $\Delta m_v = 3.2 \times E(B-V)$, we can also derive by how much the apparent magnitude of the star has been changed due to the interstellar absorption and can correct for it. Suppose we have measured for an A0V star a $B-V=0.10$ and $U-B=0.07$. Its true $B-V=0$ as we know for an A0V star. Its $E(B-V)$ is therefore $E(B-V)=0.10$ and therefore $\Delta m_v = 0.32$. If we measured $m_v = 8.32$, its true apparent magnitude would have been $m_{v0} = 8.0$ if there had not been any interstellar absorption. If we want to determine the distance to the star we can use (19.3), but we have to use, of course, the apparent magnitude which the star would have had without the interstellar absorption, i.e. m_{v0}. With $M_V = 0.5$ we find a distance modulus $m_v - M_V = 7.5$, which puts the star at a distance of about 316 pc.

On average, we find an interstellar absorption of about one magnitude in the visual for a distance of 1 kpc. The distance to the galactic center is about 8 kpc. Even without the dense dark clouds in that direction, the stars close to the center would be expected to have an average reduction in brightness of 8 magnitudes due to the average interstellar extinction. Their colors are expected to be changed by $E(B-V)=2.4$.

19.2.4 Determination of the 'true' colors of the stars

How can we determine the 'true' color line for the stars? We can, for instance, measure the colors of nearby stars. Chances are that they are nearly unreddened. There are, however, no O or B stars nearby. If there are no nearby stars of a given kind, the best thing is usually to measure stars which are above the galactic plane, if possible. Stars in the plane are usually more reddened than those above the galactic plane because the dust is confined to a rather thin disk (see Fig. 19.8). In fact, it is this observation which tells us

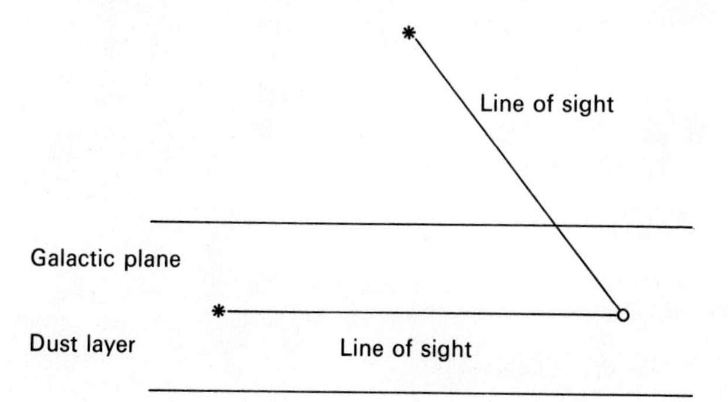

Fig. 19.8. For a given distance stars in the Galactic plane are more reddened than stars outside of the Galactic plane because the line of sight passes through the dust layer for a longer distance.

that the dust is confined to a rather thin disk. It is distributed inhomogeneously, but rarely reaches higher than about 100 pc.

Unfortunately, the O and B stars are also strongly concentrated toward the galactic plane, since they are formed from large dust clouds. So they lie entirely within the dust layer. But the dust layer is patchy and for some stars the E(B − V) is smaller than for others, so we can tentatively assume that the bluest ones have the 'true' color. For some stars there is a way to check this. Suppose we have observed a few nearby A or F stars so that we know their true colors. If the O and B stars happen to be near A or F stars, for instance, in the clusters, we can then measure the reddening for the A and F stars and hope that they will be the same for the hot O and B stars. In young clusters the extinction is, however, patchy and may be different for different stars, as, for instance, for the Pleiades, in which we can see patches of dust, seen as bright nebulosity (see Fig. 19.9), which is very inhomogeneous. The best way would be to measure colors for binaries with one O or B star and the other component being an A or F star. Unfortunately, there are not many binaries with O and B type components around for which the companion colors can also be measured, mainly because O and B stars are so short-lived, and also

Fig. 19.9. On a long-exposure photograph of the Pleiades star cluster patches of nebulosity show up in front of the stars, due to intersteller dust. This interstellar dust appears bright because it reflects the starlight in our direction.

because they are so bright that A or F stars can only be measured if they are well separated from the hotter star. We may sometimes find stars behind the O and B stars for which we may expect at least as much reddening as for the early type star. We then get an upper limit for the reddening. By putting all the available data together we can get a rather good value for the 'true' colors.

19.2.5 The wavelength dependence of the interstellar extinction

For nearby unreddened stars the whole energy distribution can be measured, for instance, by so-called scanner observations which measure the intensities in wavelength bands of 10 or 20 Å. If scanner observations are then also made for the same kind of star at a known distance, the flux ratio of the two stars, corrected for the difference in distance, tells us the interstellar extinction law, because

$$\frac{f_2(\lambda)}{f_1(\lambda)} = e^{-\tau_\lambda(\text{interstellar})}. \tag{19.13}$$

Here τ_λ (interstellar) is the optical depth of the interstellar medium between us and the star (see Fig. 19.10). $f_1(\lambda)$ is the flux of the unreddened star, and $f_2(\lambda)$ the flux of the reddened but otherwise similar star after correction for the difference in distance.

With

$$\Delta m_\lambda = -2.5 \cdot \Delta \log f_\lambda, \tag{19.14}$$

we find

$$\begin{aligned} m_\lambda(2) - m_\lambda(1) &= -2.5[\log f_2(\lambda) - \log f_1(\lambda)] \\ &= +2.5 \log e \cdot [\tau_\lambda(2) - \tau_\lambda(1)] = 2.5 \cdot \log e \cdot \tau_\lambda(2) \end{aligned} \tag{19.15}$$

because star 1 is unreddened, so $\tau_\lambda(1) = 0$.

We then obtain $\tau_\lambda(\lambda)$ for every wavelength and can measure how the interstellar extinction varies with the wavelength λ. For the wavelength band

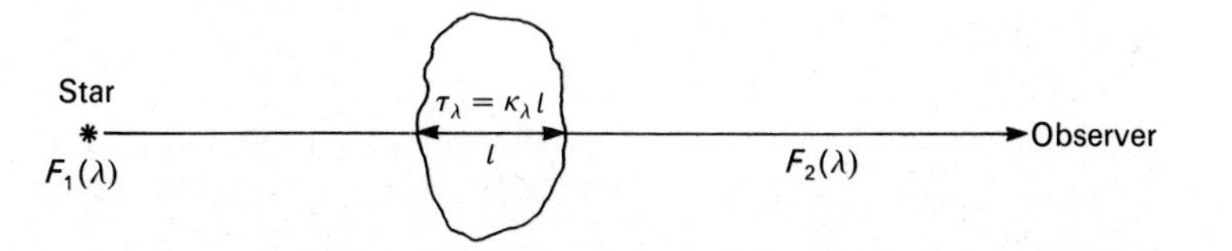

Fig. 19.10. The radiative flux $F(\lambda)$ from a star passes through an interstellar gas cloud and is reduced by the absorption in the cloud. The observed flux $F_2(\lambda) = F_1(\lambda) \cdot e^{-\tau_\lambda}$, where $F_1(\lambda)$ is the flux we would have received without the absorption in the cloud.

$3900 < \lambda < 10\,000$ Å, the wavelength region observed with most telescopes, we find nearly $\tau_\lambda \propto 1/\lambda$. For the atomic absorption of hydrogen and similar atoms or ions we find $\kappa_\lambda \propto \lambda^3$ for the absorption from each atomic level. The interstellar absorption has a very different wavelength dependence. The reason is that the interstellar absorption is mainly due to dust *grains*. The light gets scattered by these little particles. If the particles have a size comparable to the wavelength of the light, the light is not completely blocked out by the grains, but can get around the grain, although not the full amount.

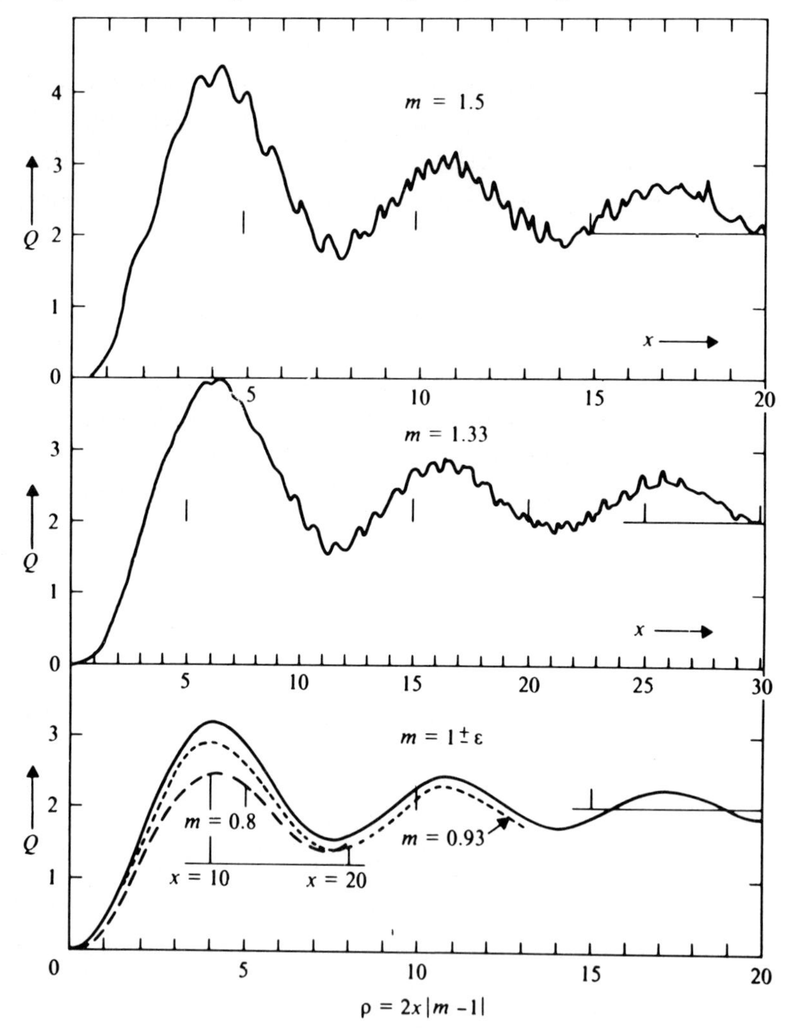

Fig. 19.11. Extinction curves for spherical grains are shown according to Mie's theory. The normalized scattering cross-section $Q = \sigma/(\pi a^2)$ is plotted as a function of $\rho = 2x|m - 1|$, where a is the radius of the dust particle, σ the actual cross-section, and x measures the ratio of the particle size to the wavelength. m is the refractive index. (From Dyson & Williams 1980.)

Think about the sound waves with wavelengths of meters which can get around the corner of a house, or the radio waves (100 m) which can also get around houses.

If one calculates the scattering on little spheres, as has been done by Mie, one finds a scattering coefficient, as seen in Fig. 19.1, where the scattering and absorption cross-section is plotted as a function of $\rho = 2x|m-1|$, where

$$x = \frac{2\pi a}{\lambda},$$

and a is the radius of the scattering particles; m is the refractive index; $|m-1|$ is of the order 0.1–0.5 for different substances. This then yields ρ to be about

$$\rho \sim \frac{a}{\lambda}. \tag{19.16}$$

The actual scattering and absorption cross-section for the particles is

$$\sigma = \pi a^2 \cdot Q, \tag{19.17}$$

such that Q gives the ratio of the scattering and absorption cross-section to the geometrical cross-section.

From Figure 19.11 we see that for $1 \leqslant \rho \leqslant 4$ the Q is approximately proportional to $\rho \sim a/\lambda$, or in other words, for $\lambda < a < 4\lambda$, or for dust particle sizes of the order of a few wavelengths of the light the extinction cross-section is roughly proportional to $1/\lambda$. If $a \approx 10^4$ Å, we expect a linear increase of the absorption cross-section σ with $1/\lambda$ for wavelengths between 10 000 and 2500 Å. When ρ becomes very small, which means for very large λ, the extinction goes to zero.

For very long wavelengths the interstellar grains do not absorb or scatter, which means that the interstellar dust clouds become transparent to radio waves. (If you put a football into the path of a sound wave, it does not influence the sound wave.) Therefore, for radio waves the Milky Way dust is transparent. Already in the infrared the extinction becomes very small, almost zero. For very short wavelengths the cross-section for absorption and scattering does not increase any more but remains, finally, about constant with wavelength. Grains much larger than the wavelength of the light simply block the light which hits them. Only the light beams passing between the grains get to us. This fraction of the light is the same for all wavelengths.

19.2.6 The ultraviolet extinction

Since we have rockets and satellites, we can also measure the extinction curves in the ultraviolet. Fig. 19.12 shows the approximate dependence on $1/\lambda$: the right-hand end of the curve corresponds to the wavelength of 1215 Å, which is the wavelength of the hydrogen Lyα line, which is the strongest hydrogen line because it corresponds to the transition of the electron from the ground level to the next higher level, namely the level with the main quantum number $n = 2$. The interstellar medium absorbs strongly in all the hydrogen Lyman lines originating from the ground level. Therefore, we cannot measure the dust extinction in the wavelength region of these lines. The overall run of the extinction curve in the ultraviolet does not look like that of the Mie curve seen in Fig. 19.11. There is a hump around 2200 Å, which is about 400 Å wide. For shorter wavelengths we find first a decrease and then again a rather steep increase of the absorption.

This behavior may be understood if we have a mixture of particles of different sizes. The smaller particles then become important for shorter wavelengths. The hump needs a special interpretation. Most astronomers believe it is due to very fine graphite particles, although this is not yet certain.

When interpreting these extinction curves, we also have to keep in mind that the Mie theory considers only spherical dust grains. We do not know how the cross-sections change for odd particle shapes.

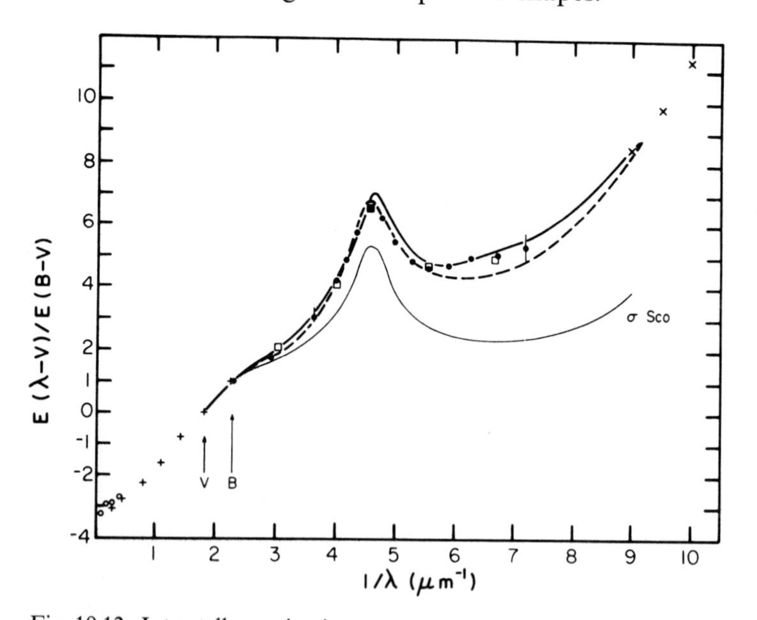

Fig. 19.12. Interstellar extinction curves, $m_\lambda - m_V$, per E(B − V) = 1, also written as $(A_\lambda - A_V)$/E(B − V), or E(λ − V)/E(B − V), are shown as a function of $1/\lambda$, where λ is the wavelength of the light measured in μ. ($1\mu = 10^4$ Å). The bump occurs at 2200 Å. The average extinction curve is shown as the dashed line. Low extinctions in the ultraviolet are relatively rare and do not influence the average very much.

The interstellar extinction for the short wavelengths does not look like Mie curves seen in Fig. 9.10. (From Dyson & Williams 1980.)

The interesting part is that this extinction curve is not the same everywhere in the sky. In regions of star formation it looks different. What changes is the size of the hump relative to the rest, and also the degree of increase for very short λ. In dense interstellar clouds, where star formation is going on, the increase at short wavelengths is much smaller and often the hump is less high. In Fig. 19.12 we compare some of these curves with the standard curve. People try to understand this by taking into account the strong ultraviolet radiation of the newly formed hot stars. They speculate that this radiation destroys the small grains, resulting in a smaller increase of the extinction for the short wavelengths.

In the external galaxy next to us, the Large Magellanic Cloud (LMC), the increase towards the ultraviolet is even steeper than shown in Fig. 19.12. The absorbing particles responsible for this increase must be relatively more abundant in the LMC than in the Galaxy. If these were the small grains we wonder why the many young, hot stars in the LMC do not destroy these small grains.

It seems that the interstellar medium in the LMC is, on average, very transparent. There seem to be very few large grains in the LMC.

19.3 The interstellar gas

The interstellar gas is of little importance for the observations of cool stars, except for the observations of the transition layer and coronal lines and the X-ray observations.

For the hot stars it is important that the hydrogen in the interstellar gas absorbs all light being emitted at wavelengths shorter than 912 Å, except for the very closest hot stars, which are, however, all white dwarfs.

Fortunately, cool atomic hydrogen gas can absorb at all wavelengths, which means continuously over all wavelengths, only for $\lambda \leqslant 912$ Å. For wavelengths shorter than 912 Å all hydrogen atoms in the ground level can contribute to this absorption. At 912 Å the absorption coefficient per hydrogen atom is about 10^{-17} cm^2. With an average density of about 1 hydrogen atom per cm^3 in the interstellar medium (with laboratory standards this is, of course, an extremely good vacuum), there are about 3×10^{19} atoms between us and a star at 10 pc distance. The optical depth τ_λ along the line of sight at 912 Å is then $\tau_\lambda \approx 300$, and the amount of radiation which we would receive is

$$f_\lambda = f_{\lambda 0} e^{-300},$$

where $f_{\lambda 0}$ is the flux which we would have received without any hydrogen absorption. This means we would not be able to measure anything.

Fortunately, there are some directions in the sky for which the hydrogen density along the line of sight is only about 10^{-3} atoms cm^{-3}. In this direction we may then still see far ultraviolet light from hot stars within a distance of about 50 pc. There are, however, no O stars and very few early B stars in this direction within 50 pc, but with an appropriate camera on a satellite those few stars could be observed at these short wavelengths, which would be very important to do.

Our estimates show that generally we cannot expect to receive any of the far ultraviolet light of the O and B stars, unless we go to extremely short wavelengths, namely the X-ray region, where the absorption coefficient of hydrogen is much smaller.

For the O stars most of the radiation is, however, emitted at the unobservable wavelengths. Therefore, we can only observe a relatively small fraction of the total light emitted by these stars. It is therefore very difficult to determine accurate effective temperatures for these stars when we see only a minor fraction of the emitted energy.

The interstellar gas is also quite important for the interpretation of X-ray observations. At X-ray wavelengths the hydrogen and helium absorption coefficients are much lower than the absorption coefficient, around 900 Å, but they are still large enough to absorb a large fraction of the X-rays emitted by hot stars and by extragalactic sources. In Fig. 19.13 we show the calculated dependence of the absorption coefficient on wavelength for the short wavelengths, including the extreme ultraviolet and the X-ray region.

In the optical spectral region the interstellar gas is seen by means of its interstellar absorption lines. Lines of neutral particles, of ionized particles, and of interstellar molecules are seen. In the interstellar medium collisions are very rare. The radiation density usually is also quite low. There are, therefore, very few processes to excite the atoms or ions to higher states of excitation. The interstellar lines therefore usually are due to absorption processes from the ground state of the atoms or ions. Only in very dense regions may we see lines originating from states with very low but non-zero excitation energies. Interstellar absorption lines can usually be recognized easily because only absorption lines from the ground state are seen. In addition, the interstellar lines in the optical region are generally quite sharp. For rapidly rotating hot stars, they can easily be recognized because of this. In Fig. 19.14 we show several spectra in which the interstellar absorption lines from the gas can be seen as sharp lines, while the stellar absorption lines are broadened by rotation.

For these interstellar absorption lines we can also measure the line strengths from which we can determine the number of absorbing atoms along the line of sight. If we do that for all the lines which can be observed in the optical and in the ultraviolet spectral region, we can determine the relative abundances of the elements along the line of sight. The elements are mostly singly ionized. (Once an ultraviolet photon ionizes an atom, the remaining ion finds it hard to meet another electron with which to recombine because the densities are so low in the interstellar medium.) If we want to study relative abundances, we usually have to look at the lines of the singly ionized ions. There are exceptions, though, for atoms which have rather large ionization energy, like nitrogen or oxygen, and which would need far ultraviolet photons for ionization. These are mainly neutral. Such abundance analysis and the comparison with the abundance of hydrogen

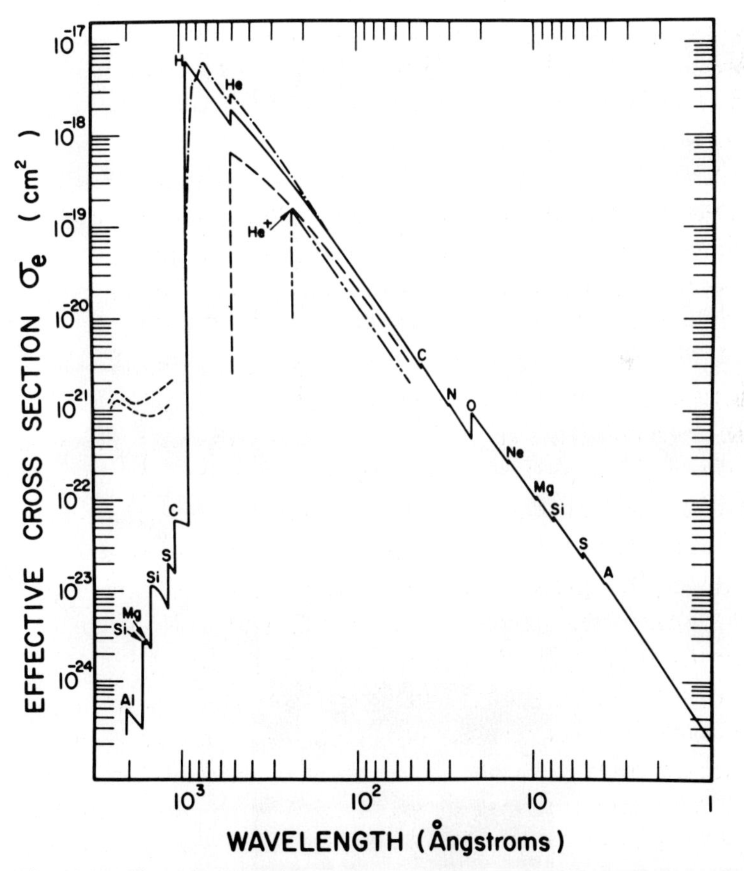

Fig. 19.13. The calculated interstellar extinction is shown also for wavelengths between 100 and 912 Å, which generally cannot be observed. For wavelengths shorter than about 20 Å the extinction becomes smaller than the dust extinction in the visual spectral region. Therefore, X-rays can penetrate the interstellar medium, while radiation with wavelengths between about 100 and 900 Å generally cannot. (From Cruddace *et al.* 1974.)

showed that most of the heavy elements are depleted in the interstellar medium with respect to hydrogen as compared to the solar material.

Obviously, most of the heavy elements are locked up in grains. Clearly, grains cannot form from hydrogen gas. Only a small amount of mass in the form of grains can lock up all the heavy atoms because there are so few of them to begin with. If we want to determine the number of hydrogen atoms absorbing at λ 912 Å from the observed abundance of, say, Ca^+ along the line of sight using the abundance ratio of Ca to hydrogen, we have to take this into account.

For all stellar observations we have to be aware of the effects of interstellar absorption, which may vary considerably from one place to another.

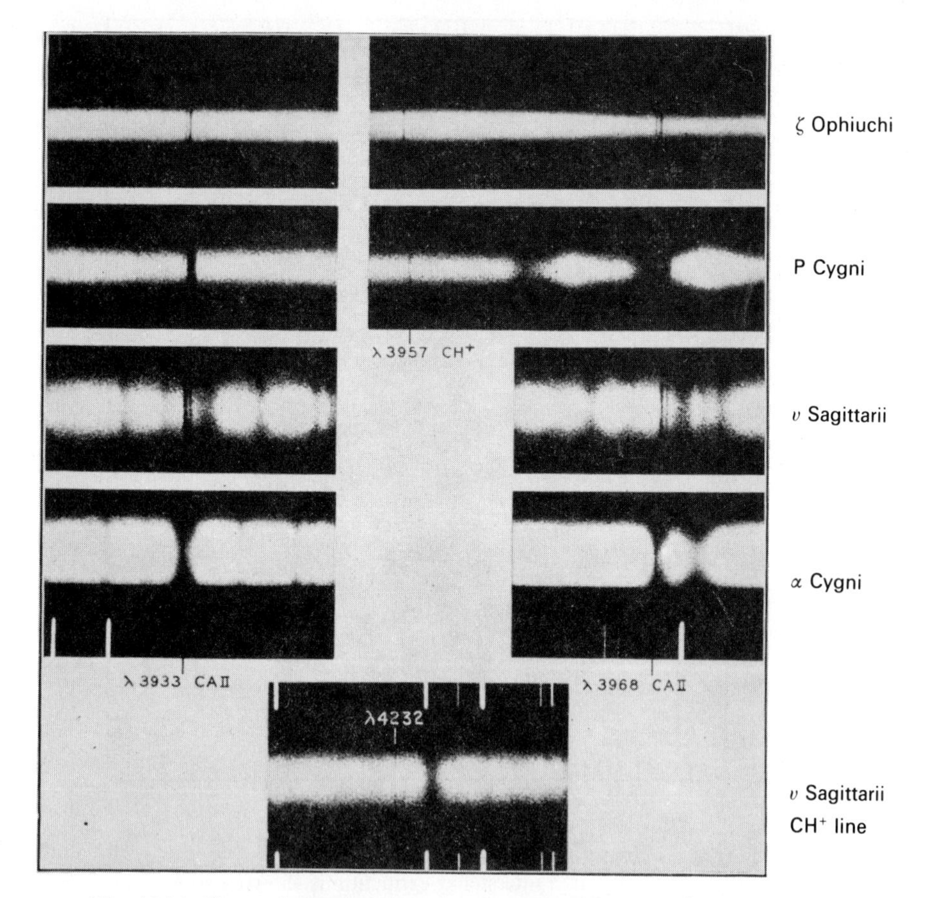

Fig. 19.14. Sharp absorption lines due to absorption by Ca^+ ions in the interstellar gas are seen in the spectra of the stars ζ Ophiuchi (top), P Cygni, υ Saggitarii and α Cygni. In the spectrum of υ Sagittarii a sharp line due to the ionized molecule CH^+ (present also in the interstellar medium) is also seen. (From Greenstein 1951.)

Appendix A

Problems

Chapter 1

1. The altitude of a star is the angular height of the star above the horizon; it changes during the day and night because of the Earth's rotation. What is the largest altitude which the stars can reach for an observer at a latitude of 35° if the declinations of the stars are as follows:

Star	1	2	3
δ	$-22°$	$+30°$	$+90°$

2. One siderial day is the true rotation period of the Earth, as can be seen from the apparent rotation of the sky. There are about 366.25 siderial days in a year. We experience one day less per year because the Earth orbits the sun once a year. A star with right ascension $\alpha = 8^h.00$ is in the meridian at 8:00 hours siderial time. At this time it has the greatest altitude. The difference between siderial time and right ascension is called the hour angle. For the star with $\delta = -22°$ plot the altitude as a function of the hour angle for an observer at 35° latitude. For how many hours is the star above the horizon?

3. Why can we see some stars only during the winter and some only during the summer?

Chapter 2

1. A star has a proper motion of $\mu_\delta = 0.1$ arcsec and $\mu_\alpha = 0$ per year. (μ_α is the proper motion in the direction of right ascension, μ_δ in the direction of declination.) The star has a radial velocity $v_r = 20$ km/s. What is its velocity relative to us, if it is at (a) a distance of 20 pc? and (b) at a distance of 100 pc?

2. Another star has a declination of $\delta = +60°$. It has $\mu_\alpha = 0.1$ sec and $\mu_\delta = 0.5$ arcsec per year, $v_r = 20$ km/s. How large is its total velocity relative to us if it is at a distance of 20 pc?

Chapter 3

1. You want to measure the U, B, V colors of an unknown star. You have a photometer whose counts per minute are proportional to the fluxes which you receive. You measure the standard star, say Vega (you can of course also take another star whose apparent magnitudes you know accurately) for 1 minute each at different zenith distances z, and obtain the following counts for the different filters:

z	V	B	U
20°	10 000	8000	5000
40°	9 460	7247	4223
60°	8 063	5451	2597

Determine τ_V, τ_B, τ_U for airmass 1.

2. After you have determined the extinction correction for the Earth's atmosphere you measure the fluxes for the unknown star. It is much fainter than your standard star, therefore you integrate for 10 minutes each. You measure the following counts:

	V	B	U
$z = 40°$	3784	2898	1689

Determine the m_V, m_B, m_U.
Which kind of a star is it if it is a main sequence star?

3. For a star with $m_V = 3.0$ the trigonometric parallax was measured to be 0.03 arcsec. What is the distance of this star? What is its absolute magnitude M_V?

Chapter 5

1. For an unknown star with $m_V = 5.5$ you have measured the colors and found $(U - B)_0 = 0.0$, $(B - V)_0 = 0.0$. Assuming it is a main sequence star, at what distance is it? What should be its parallax angle π? Could it be measured?

For another star you measured $(U - B)_0 = 0.12$, $(B - V)_0 = 0.63$, $m_V = 7.75$, at what distance is this star if it is a main sequence star?

2. Suppose you were to observe a visual binary system. You find that both stars have $(U - B)_0 = 0.12$ and $(B - V)_0 = 0.63$. The magnitude of the fainter star is $m_V = 8.5$ and the brighter star has $m_V = 5.5$. Which kind of a star is the brighter one? What is the distance modulus of these stars?

Chapter 6

1. Vega has $M_V = 0.5$. How large is M_{bol}?

Calculate the luminosity of Vega. Remember that $M_{bol\odot} = 4.75$. If Vega has a mass of 2.7 $M_\odot$ what is the ratio of L/M for Vega? Compare with the solar value.

2. After correction for interstellar extinction the star τ Scorpii has $B - V = -0.30$. Its $M_V = -4$. How large is its M_{bol}? Use $BC = 3.0$. Calculate its luminosity L. If its mass is 15 $M_\odot$ what is its ratio of L/M? Make a plot of L/M for the sun, Vega and τ Sco. If we approximate the relation by $L \propto M^\beta$ how large is β approximately?

Chapter 8

1. For the star α Orionis an angular diameter of 0.048 arcsec was measured with the Michelson interferometer. Its average visual magnitude is $\bar{m}_V = 0.8$. The bolometric correction for this star is about $BC = 1.50$. What is the average effective temperature of α Orionis? Remember that for the sun $m_{bol} = -26.85$, and the solar constant is $S = 1.38 \times 10^6 \text{ erg cm}^{-2} \text{ s}^{-1}$.

2. For Vega an angular diameter of 3.24×10^{-3} arcsec was measured by Hanbury Brown. The bolometric correction for Vega is $BC = 0.15$. What is the effective temperature of Vega?

The parallax angle of Vega was measured to be $\pi = 0.123$ arcsec. What is the radius of Vega? Use these values of R and T_{eff} to calculate the luminosity of Vega and its absolute bolometric magnitude. Compare this value with the value obtained from its M_V and the bolometric correction. How could the difference perhaps be explained?

3. For Sirius, Hanbury Brown measured an angular *diameter* $\theta = 5.89 \times 10^{-3}$ arcsec. The apparent visual magnitude of Sirius is $m_V = -1.46$. The bolometric correction is $BC \approx 0.15$. What is the T_{eff} of Sirius? Remember that for the sun $m_{bol} = -26.85$, and the solar constant $S = 1.38. \times 10^6 \text{ erg cm}^2 \text{ s}^{-1}$.

4. Sirius has a faint companion, Sirius B, which has an apparent visual magnitude $m_V = 8.68$. Sirius is at a distance of 2.65 pc. Calculate the absolute visual magnitude of Sirius B.

The energy distribution of this star shows that it has $T_{eff} = 26\,000$ K. The bolometric correction for this temperature is $BC = 2.75$. For the sun $BC = 0.07$. Calculate the M_{bol} for Sirius B. Calculate the luminosity of this star and its radius. Which kind of a star is this? Remember $L_\odot \approx 4 \times 10^{33} \text{ erg s}^{-1}$.

From the binary orbits we determine that the mass of Sirius B is M (Sirius B) $\sim 1\,M_\odot$. Calculate the average density of Sirius B.

Chapter 9

1. A spectral line whose laboratory wavelength is $\lambda_0 = 4343.0$ Å is seen in a stellar spectrum at a wavelength of 4344.2 Å. What is the line of sight velocity component of the star relative to us? Is the star moving away from us or towards us?

2. For a binary system the ratio of the radial velocities for the two stars was measured to be $v_r(1)/v_r(2) = 1.5$. One of the stars is a solar type star, which means a G2V star. What is the mass of the other star?

3. For an eclipsing binary with a period of 30 days the total duration of the eclipse is 8 hours. We see a flat minimum for 1 hour and 18 minutes. The radial velocity amplitude for star 1 is $v_{r1} = 30 \text{ km s}^{-1}$ and for star 2 it is $v_{r2} = 40 \text{ km s}^{-1}$. For a circular orbit and $i = 90°$ how large are the radii of the stars? Estimate the masses of the stars.

Chapter 10

1. In Fig. A.1 we show low resolution stellar spectra of some stars. The wavelengths for a few strong lines are given. Estimate the spectral types of these stars.

Chapter 11

1. In stellar spectra we see mainly absorption lines. What does this tell us about the temperature stratification in stars?

2. In the solar spectrum we see a strong line at 4303.177 Å. We suspect that this is an FeII line, i.e. a line from the Fe^+ ion. Check from the Multiplet tables of Mrs Moore Sitterly which other FeII lines should be present if this line identification is correct. Which lines should be stronger and which lines should be weaker than the line at 4303.177 Å?

Chapter 13

1. For A stars we usually observe very broad lines due to rotational broadening. We call the equatorial rotational velocity v_{rot}. If for a given star the full line width from edge to edge is 0.8 Å for a line at 4343 Å, how large is $v_{rot} \sin i$?

For magnetic stars we can measure the rotation period. Suppose that for the observed star a rotation period of 3 days is found, how large is $\sin i$?

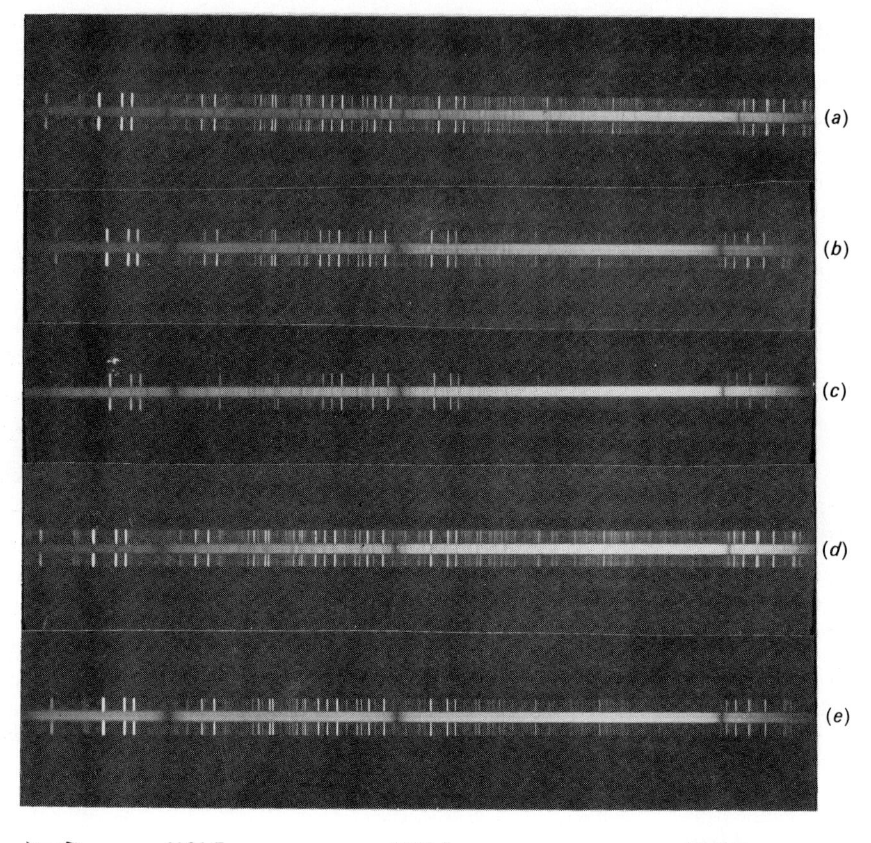

Fig. A1. A series of stellar spectra showing dark absorption lines in a bright continuum. The wavelengths of the strongest lines are given at the bottom in Å. Laboratory emission line spectra are seen on both sides of the stellar spectra. They are used to determined the wavelengths of the stellar absorption lines.

Chapter 14

1. Suppose that for a magnetic star we observe for a normal Lorentz triplet a shift of the two circularly polarized components by 0.2 Å from the central component. How large is the longitudinal magnetic field component?

2. For an average peculiar A star $v_{rot} \sin i$ is about 70 km s^{-1}. How large a magnetic field would such a star need to have before we could measure the splitting of the Lorentz triplet for a line at 4500 Å?

Chapter 16

1. A δ Cephei star in the Galaxy has $\bar{m}_V = 3.7$ and a pulsation period of 5.6 days. How far away is it?

2. In the Andromeda Galaxy which has a distance modulus of $m_V - M_V = 24.1$ we can now just barely observe RR Lyrae stars which have an average absolute magnitude of $M_V = 0.6$. What are the apparent visual magnitudes of these stars in the Andromeda galaxy?

3. The RR Lyrae stars in the globular cluster M3 have an apparent visual magnitude of $m_V = 15.6$. How far away is the globular cluster? If we take the plane of our Galaxy as the reference plane then M3 has an altitude of 77° above this plane. We call this angle the galactic latitude b. How high, above the Galactic plane is M3, which means how large is z in Fig. A.2 below.

Chapter 17

1. If the supernova 1987A in the Large Magellanic Cloud keeps expanding with its present velocity of $10\,000\,\mathrm{km\,s^{-1}}$ for another year, how large would it be in pc after 10 years? How large would it then be in seconds of arc? The presently adopted distance modulus to the LMC is $m_V - M_V = 18.3$.

2. If in the near future one of the Pleiades stars should become a supernova with the same intrinsic brightness as the supernova 1987A in the LMC which reached $m_V = 2.9$, what would be the apparent brightness of such a Pleiades supernova? If it should also expand like the LMC supernova at $10\,000\,\mathrm{km\,s^{-1}}$ how large would it be after 10 years as measured in arcsec? Compare the size and brightness with the moon.

Chapter 18

1. In the solar granulation we see temperature differences of up to about 400 K. If the effective temperatures of a dark and a bright spot are 5600 K and 6000 K respectively what is the intensity ratio of the two spots at 5000 Å if they radiate like black bodies?

2. In the solar granulation we see velocity differences of about $2\,\mathrm{km\,s^{-1}}$. How large is the Doppler shift for such velocities for a wavelength of 5000 Å? Compare this with the rotational broadening of the solar spectral lines. From the motion of the sunspots we determine that the rotational period for the sun is 27 days.

Chapter 19

1. For a main sequence star of spectral type B we measure $B - V = 0.3$, $U - B = -0.36$. What are its true colors? How large is its color excess? Its apparent magnitude is $m_V = 5.0$. What would be its m_{V_0}, m_{B_0}, m_{U_0} magnitudes, and its $(B - V)_0$ and $(U - B)_0$ colors? Use $R = \Delta m_V / E(B - V) = 3.2$. How far away is this star?

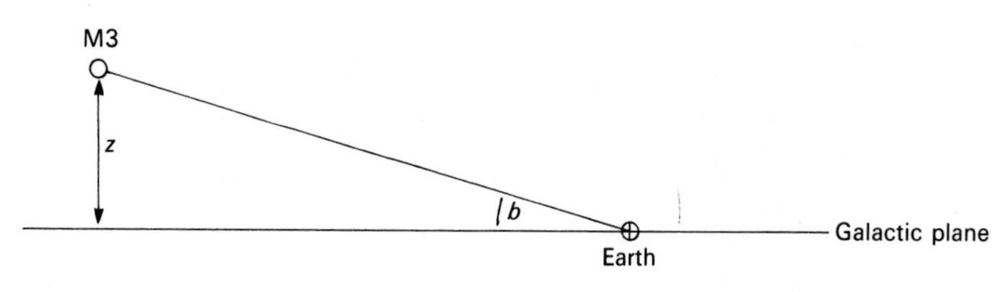

Fig. A2

Appendix B

Some important astronomical quantities

I. Atomic quantities

Electron mass $m = 9.105 \times 10^{-28}$ g
Proton mass $m_H = 1.672 \times 10^{-24}$ g
Planck's constant $h = 6.624 \times 10^{-27}$ erg s
Elementary charge $e = 4.802 \times 10^{-10}$ esu
Velocity of light $c = 2.998 \times 10^{10}$ cm s^{-1}
Stefan–Boltzmann constant $\sigma = 5.672 \times 10^{-5}$ erg cm^{-2} s^{-1} K^{-4}
Boltzmann constant $k = 1.38 \times 10^{-16}$ erg K^{-1}
Gas constant $R_g = 8.314 \times 10^{7}$ erg K^{-1} mol^{-1}

$\chi_{\text{ion}}(\text{H}) = 13.60$ eV
$\chi_{\text{ion}}(\text{He}) = 24.58$ eV
$\chi_{\text{ion}}(\text{He}^{+}) = 54.4$ eV
$\chi_{\text{ion}}(\text{Fe}) = 7.90$ eV
$\chi_{\text{ion}}(\text{Si}) = 8.15$ eV
$\chi_{\text{ion}}(\text{H}^{-}) = 0.7$ eV

Absorption edges for Hydrogen continua

Lyman continuum: 912 Å
Balmer continuum: 3647 Å
Paschen continuum: 8208 Å
H^{-} continuum: 17 000 Å
Helium Lyman continuum: 504 Å
Helium^{+} Lyman continuum: 228 Å

Hydrogen lines

Lyα: 1215.67 Å
Lyβ: 1025.72 Å
Lyγ: 972.5 Å
Hα: 6562.82 Å
Hβ: 4861.33 Å
Hγ: 4340.47 Å
Hδ: 4101.74 Å

Some important resonance lines of other elements

He = He I	$\lambda = 591.42$ Å
He^{+} = He II	303.78 Å
Ca = Ca I	4226.73 Å

$Ca^+ = Ca$ II	3933.66; 3968.47 Å
$Na = Na$ I	5889.95; 5895.92 Å
$C = C$ I	1657.00 Å
$C^+ = C$ II	1335.68; 1334.52 Å
$C^{+++} = C$ IV	1548.20; 1550.77 Å
$O = O$ I	1302.17; 1304.86; 1306.02 Å
$K = K$ I	7664.91; 7698.98 Å
$Ba = Ba$ I	5535.48 Å
$Ba^+ = Ba$ II	4554.03 Å
$Sr = Sr$ I	4607.33 Å
$Sr^+ = Sr$ II	4077.71 Å
$Mg = Mg$ I	4571.10; 2852.12 Å
$Mg^+ = Mg$ II	2795.52; 2802.70 Å

II. Earth

1 Siderial day = rotation period of the Earth: $23^h56^m04^s.1$
1 year = 365.256 solar days
Inclination of equatorial plane to ecliptic: 23°27′
1 astronomical unti 1 au = 1.496×10^{13} cm
Mass of the earth $M_\oplus = 5.98 \times 10^{27}$ g
Equatorial radius $R_\oplus = 6.371 \times 10^7$ cm = 6371 km

III. Sun

Rotation Period $P_\odot = 25.38$ days
Mass $M_\odot = 1.98 \times 10^{33}$ g
Radius $R_\odot = 6.96 \times 10^{10}$ cm
Gravitational acceleration $g_\odot = 2.74 \times 10^4$ cm s^{-2}
Mean density $\bar{\rho} = 1.41$ g
Angular radius $R/d = 959.63$ arcsec = 0.0046526 radians
Solar constant $S = 1.38 \times 10^6$ erg cm^{-2} s^{-1}
Luminosity $L_\odot = 3.96 \times 10^{33}$ erg s^{-1}
Surface flux $\pi F_\odot = 6.41 \times 10^6$ erg cm^{-2} s^{-1}
Effective temperature $T_{\rm eff} = 5800 \pm 20$ K
Apparent visual magnitude $m_{V\odot} = -26.78$
Apparent bolometric magnitude $m_{\rm bol,\odot} = -26.85$
$B - V = 0.63 \pm 0.02$
$U - B = 0.13 \pm 0.02$
Absolute visual magnitude $M_{V\odot} = 4.82$
Absolute bolometric magnitude $M_{\rm bol,\odot} = 4.75$

IV. Some other important quantities

1 eV = 1.602×10^{-12} erg
1 parsec = 3.2598 light years = 3.084×10^{18} cm
gravitational constant $G = 6.67 \times 10^{-8}$ dyn cm^2 g^{-2}
0° Kelvin = −273.16 centigrade
$m_V = 0$ corresponds to $\pi f_\lambda = 3.6 \times 10^{-9}$ erg cm^{-1} s^{-1}/Å
or $\pi f_\lambda = 3.6 \times 10^{-16}$ watt cm^{-2}/Å at 5500 Å.

References

I. Other textbooks

All introductory astronomy textbooks.

Also:

Dufay, J. (1964). *Introduction to Astrophysics: The Stars* (Translated by O. Gingerich). Dover Publications, New York.
Novotny, E. (1973). *Introduction to Stellar Atmospheres and Interiors*. Oxford University Press, Oxford.
Shu, F. (1982). *The Physical Universe*. University Science Books, Mill Valley, California.
Swihart, T. L. (1968). *Astrophysics and Stellar Astronomy*. John Wiley and Sons, Inc. New York.
Unsöld, A. & Bascheck, B. (1983). *The New Cosmos*, 3rd ed. Springer Verlag, Berlin.
Voigt, H. H. (1975). *Abriss der Astronomie*, Wissenschafts Verlag, Bibliographisches Institut, Mannheim.
Walker, G. (1987). *Astronomical Observations*. Cambridge University Press, Cambridge.

II. Some Star catalogues

Bonner Durchmusterung (BD), 4th ed. (1968). Universitats Sternwarte Bonn, Verlag F. Dümmlers.
Henry Draper Catalogue, A. J. Cannon & Pickering, E. C. (1918–24). *Annals of the Astronomical Observatory of Harvard College*, Vol. 91–99.
Smithsonian Astrophysical Observatory Star Catalogue (SAO) (1966). Positions and proper motions of 258, 997 stars for the epoch and equinox of 1950.0. U.S. Government Printing Office.
The Bright Star Catalogue (BS or HR), 3rd ed., Hoffleit, D. & Jascheck, C. (1982). Yale University Observatory, New Haven.
General Catalogue of Variable Stars, Kukarkin *et al.* (1969 ff.). Astronomical Council of the Academy of Sciences, Moscow. 2nd edition: Kholopov *et al.* 1985 ff, NAUKA Publishing House, Moscow.

III. Some Reference books for stellar data

Allen, C. W. (1968). *Astrophysical Quantities*. 3rd ed. University of London. The Athlon Press.
Landolt Börnstein (1982). *Numerical Data and Functional Relationships in Science and Technology*, Group VI, Vol. 2, *Astronomy and Astrophysics*, ed. K. Schaifers and H. H. Voigt. Springer Verlag, Berlin.

Blanco, V. M., Demers, S., Douglass, G. G. & Fitzgerald, M. P. (1968). *Photoelectric Catalogue: Magnitudes and Colors of Stars in the UBV and U_c, B, V systems.* Publications of the United States Naval Observatory, 2nd Series, Vol. 21.
Jamar, C., Maceau Hercot, D., Monfils, A., Thompson, G. I., Houziaux, L. & Wilson, R. (1976). *Ultraviolet Bright Star Spectrophotometric Catalogue.* European Space Agency SR 27, SR 28.
Moore, C. E. (1959). National Bureau of Standards, Technical Note No. 36 (data on spectral lines). U.S. Government Printing Office.
Strand, K. Aa. (1963). *Basic Astronomical Data, Vol. III of Stars and Stellar Systems.* Gen. ed. G. P. Kuiper & B. M. Middlehurst. The University of Chicago Press, Chicago.
Thompson, G. I., Nandy, K., Jamar, C., Monfils, A., Houziaux, L., Carnochan, D. C. & Wilson, R. (1978). *Catalogue of Stellar Ultraviolet Fluxes.* The Science Research Council.

IV. References Quoted in the Text

Abell, G. O. (1982). *Exploration of the Universe*, 4th edn. Saunders College Publishers, Philadelphia.
Allen, C. W. (1968). *Astrophysical Quantities*, 2nd edition. University of London, the Athlon Press.
Arp, H. C. (1958). In *Handbook of Physics*, Vol. 51, p. 75. Springer Verlag, Berlin.
Babcock, H. (1960). In *Stellar Atmospheres*, ed. J. Greenstein. Volume VI of *Stars and Stellar Systems*, Gen. ed. G. P. Kuiper & B. M. Middlehurst, p. 282. The University of Chicago Press, Chicago.
Becker, W. (1950). *Sterne und Stern Systeme*, 2nd edition. Theodor Steinkopff, Dresden.
Biermann, L. (1941). Vierteljahresschrift, d. *Astronomischen Gesellschaft*, **76**, 194.
Bidelman, W. P. & Keenan, P. C. (1951). *Astrophysical Journal*, **114**, 473.
Binnendijk, L. (1960). *Properties of Double Stars.* University of Pennsylvania Press, Philadelphia.
Bohlender, D. A., Brown, D. N., Landstreet, J. D. & Thompson, I. B. (1987). *Astrophysical Journal*, **323**, 325.
Böhm-Vitense, E. (1984). *Science*, **223**, 777.
Böhm-Vitense, E. & Canterna, R. (1974). *Astrophysical Journal*, **194**, 629.
Böhm-Vitense, E. & Johnson, P. (1977). *Astrophysical Journal Supplement Series*, **35**, 461.
Borra, E. F., Landstreet, J. D. & Thompson, I. B. (1983). *Astrophysical Journal Supplement Series*, **53**, 151.
Bray, R. J. & Loughhead, R. E. (1974). *The Solar Chromosphere*, p. 224. Chapman & Hall, London.
Burnham, R. (1978). *Burnham's Celestial Handbook*, Vol. II & Vol. III. Dover Publ. Inc., New York.
Clayton, D. D. (1983). *Principles of Stellar Evolution and Nucleosynthesis.* The University of Chicago Press, Chicago.
Cohen, M. & Kuhi, L. V. (1979). *Astrophysical Journal Supplement*, **41**, 743.
Cox, J. P. (1980). *Theory of Stellar Pulsations.* Princeton University Press, Princeton, New Jersey.
Cruddace, R., Paresce, F., Bowyer, S. & Lampton, L. (1974). *Astrophysical Journal*, **187**, 500.
Deutsch, A. J. (1958). In *Handbook of Physics*, Vol. 51, p. 689. Springer Verlag, Berlin.
Deutsch, A. J. (1952). *Transaction of the International Astronomical Union*, **8**, 801.

Dyson, J. E. & Williams, D. A. (1980). *The Physics of the Interstellar Medium.* John Wiley & Sons, New York.
Eddington, A. S. (1926). *The Internal Constitution of Stars.* Dover Publication 1950, p. 277.
Gibson, E. G. (1973). *The Quiet Sun.* NASA, Washington, D.C.
Greenstein, J. (1951). In *Astrophysics* ed. J. A. Hynek, p. 526. McGraw Hill Book Company, Inc., New York.
Greenstein, J. (1986). *Sky and Telescope*, **72**, 349.
Hanbury Brown, R. (1968). In *Annual Review of Astronomy and Astrophysics*, **6**, 13.
Hanbury Brown, R. (1974). *The Intensity Interferometer.* Taylor & Francis Ltd. London.
Herbig, G. H. (1969). In *Nonperiodic Phenomena in Variable Stars*, ed. L. Detre, p. 75. Dordrecht: Reidel.
Herbig, G. H. & Jones, B. F. (1981). *Astronomical Journal*, **86**, 1232.
Johnson, H. L. & Morgan, W. W. (1953). *Astrophysical Journal*, **117**, 313.
Joy, A. H. (1960). In *Stellar Atmospheres*, ed. J. Greenstein. Vol. VI of *Stars and Stellar Systems*, Gen. ed. G. P. Kuiper & B. M. Middlehurst, p. 653. The University of Chicago Press, Chicago.
Kiepenheuer, K. O. (1953). In *The Sun*, ed. G. P. Kuiper, p. 322. The University of Chicago Press, Chicago.
Lamla, E. (1982). In *Landolt–Börnstein: Numerical Data and Functional Relationships in Science and Technology. New Series Volume 2, Astronomy and Astrophysics* ed. K. Schaifers and H. H. Voigt, p. 35. Springer Verlag, Berlin.
Ledoux, P. & Walraven, Th. (1958). In *Handbook of Physics*, Vol. 51, p. 353. Springer Verlag, Berlin.
Morgan, W. W., Keenan, P. C. & Kellman, E. (1943). *An Atlas of Stellar Spectra.* Astrophysical Monographs, Chicago.
Morgan, W. W., Abt, H. A. & Tapscott, J. W. (1978). *Revised MK Spectral Atlas for Stars Earlier than the Sun.* Yerkes Observatory, University of Chicago and Kitt Peak National Observatory.
Paczinski, B. (1985). In *Astronomy and Astrophysics*, ed. M. S. Roberts, p. 143. The American Association for the Advancement of Science.
Parker, E. N. (1974). *Solar Physics*, **36**, 249.
Pettit, E. (1951). In *Astrophysics, a Topical Symposium*, ed. J. A. Hynek, p. 259. McGraw-Hill Book Company, New York.
Popper, D. (1980). *Annual Review of Astronomy and Astrophysics*, **18**, 115.
Ridgway, S. T., Wells, D. C. & Joyce, R. R. (1977). *Astronomical Journal*, **82**, 414.
Sandage, A. & Walter, M. (1966). *Astrophysical Journal*, **143**, 313.
Schmidt-Kaler, Th. (1982). In *Landolt-Börnstein: Numerical Data and Functional Relationships in Science and Technology, New Series, Volume 2, Astronomy and Astrophysics*, ed. K. Schaifers & H. H. Voigt, p. 1. Springer Verlag, Berlin.
Schwartz, R. D. (1983). *Annual Review Astronomy and Astrophysics*, **21**, 209.
Shu, F. H. (1982). *The Physical Universe.* University Science Books, Mill Valley, Cal.
Smart, W. M. (1977). *Textbook on Spherical Astronomy.* 6th ed., Cambridge University Press, Cambridge.
Unsöld, A. (1938). *Physik d. Sternatmosphären.* Springer Verlag, Berlin.
Unsöld, A. (1982). *The New Cosmos.* Springer Verlag, Berlin.
Zirin, H. (1966). *The Solar Atmosphere*, p. 282. Blaisdell Publishing Company, Waltham Massachusetts.
Zwicky, F. (1965). In *Stellar Structure*, ed. L. H. Aller & D. B. McLaughlin, p. 367, Vol. VIII of *Stars and Stellar Systems*, Gen. ed. G. P. Kuiper and B. M. Middlehurst. University of Chicago Press, Chicago.

Index (Volume 1)

Page numbers printed in *italic* type indicate the main source of information